SpringerBriefs in Education

We are delighted to announce SpringerBriefs in Education, an innovative product type that combines elements of both journals and books. Briefs present concise summaries of cutting-edge research and practical applications in education. Featuring compact volumes of 50 to 125 pages, the SpringerBriefs in Education allow authors to present their ideas and readers to absorb them with a minimal time investment. Briefs are published as part of Springer's eBook Collection. In addition, Briefs are available for individual print and electronic purchase.

SpringerBriefs in Education cover a broad range of educational fields such as: Science Education, Higher Education, Educational Psychology, Assessment & Evaluation, Language Education, Mathematics Education, Educational Technology, Medical Education and Educational Policy.

SpringerBriefs typically offer an outlet for:

- An introduction to a (sub)field in education summarizing and giving an overview of theories, issues, core concepts and/or key literature in a particular field
- A timely report of state-of-the art analytical techniques and instruments in the field of educational research
- A presentation of core educational concepts
- An overview of a testing and evaluation method
- A snapshot of a hot or emerging topic or policy change
- An in-depth case study
- A literature review
- A report/review study of a survey
- An elaborated thesis

Both solicited and unsolicited manuscripts are considered for publication in the SpringerBriefs in Education series. Potential authors are warmly invited to complete and submit the Briefs Author Proposal form. All projects will be submitted to editorial review by editorial advisors.

SpringerBriefs are characterized by expedited production schedules with the aim for publication 8 to 12 weeks after acceptance and fast, global electronic dissemination through our online platform SpringerLink. The standard concise author contracts guarantee that:

- an individual ISBN is assigned to each manuscript
- each manuscript is copyrighted in the name of the author
- the author retains the right to post the pre-publication version on his/her website or that of his/her institution

Margaret S. Barrett · Heidi M. Westerlund

Music Education, Ecopolitical Professionalism, and Public Pedagogy

Towards Systems Transformation

Margaret S. Barrett
Sir Zelman Cowen School of Music
and Performance
Monash University
Melbourne, Victoria, Australia

Heidi M. Westerlund
Sibelius Academy
University of the Arts Helsinki
Helsinki, Finland

ISSN 2211-1921　　　　　　　ISSN 2211-193X (electronic)
SpringerBriefs in Education
ISBN 978-3-031-45892-7　　　ISBN 978-3-031-45893-4 (eBook)
https://doi.org/10.1007/978-3-031-45893-4

© The Author(s), under exclusive license to Springer Nature Switzerland AG 2023

This work is subject to copyright. All rights are solely and exclusively licensed by the Publisher, whether the whole or part of the material is concerned, specifically the rights of translation, reprinting, reuse of illustrations, recitation, broadcasting, reproduction on microfilms or in any other physical way, and transmission or information storage and retrieval, electronic adaptation, computer software, or by similar or dissimilar methodology now known or hereafter developed.

The use of general descriptive names, registered names, trademarks, service marks, etc. in this publication does not imply, even in the absence of a specific statement, that such names are exempt from the relevant protective laws and regulations and therefore free for general use.

The publisher, the authors, and the editors are safe to assume that the advice and information in this book are believed to be true and accurate at the date of publication. Neither the publisher nor the authors or the editors give a warranty, expressed or implied, with respect to the material contained herein or for any errors or omissions that may have been made. The publisher remains neutral with regard to jurisdictional claims in published maps and institutional affiliations.

This Springer imprint is published by the registered company Springer Nature Switzerland AG
The registered company address is: Gewerbestrasse 11, 6330 Cham, Switzerland

Paper in this product is recyclable.

Acknowledgements

We are indebted to the individuals who have shared their life stories in this book: Tuulikki Laes, Riju Tulahadur, and Ricky Kej. This research work was supported by the project Music Education, Professionalism, and Eco-Politics (EcoPolitics), funded by the Academy of Finland [Grant number 338952] and Monash University (Faculty of Arts Research Support). We thank also Sandra Stauffer, Sidsel Karlsen, and Albi Odendaal and the three anonymous reviewers for their valuable comments on the draft of this book.

All data were generated in compliance with the ethical requirements of Monash University Human Research Ethics Committee (Application No. 26669 approved on October 26, 2020).

Contents

1	**Introduction**	1
	Professional Responsibility as an Imaginative Possibility	4
	A Shift from the Ego-Logical to the Eco-Logical	6
	The Potential of Public Pedagogy	8
	A Note on the Research Approach	10
	References	11
2	**The Emerging Ecological Shift**	15
	Working "Across Scales"	16
	Ecological Thinking Beyond Sustainability	17
	Towards *Eco*-Logical Thinking in Music and Music Education	20
	Concluding Remarks	22
	References	22
3	**Rethinking Professionalism in Music Education**	25
	Developing Ecopolitical Music Education Professionalism Within—Beyond Technical Rationality	25
	Music Educators as Transformative Systems Practitioners	30
	Expanding Moral Ecologies of Music Education Through Public Pedagogy	34
	Concluding Remarks	37
	References	37
4	**Using Moral Imagination**	41
	Musical Beginnings	41
	Music as a "Lifestyle"	42
	Stepping off the Path	42
	Studying Music Education	43
	Recognising One's Strengths	44
	Learning to Trust Relationality	44

	Generating Systems Transformation	45
	A Call to Research	45
	A Researcher in Her Own Terms: An Academic Entrepreneur	46
	Living Relational Values	47
	Taking Advantage of Serendipitous Moments	48
	Working Across Boundaries	48
	Changing Perceptions, Beliefs, and Key Goals	49
	Showing not Telling	50
	Hiking the Horizontal	51
	References	53
5	**Ecopolitical Systems Reflexivity in Practice**	**55**
	Professional Life of a Musician in a Land of Festivals	55
	Music Learning in Diverse Musical Ecosystems	56
	Professional Learning Through Teaching	56
	Realising One's Place in Society	57
	Musician for Musician Project: Combining Humanitarian Work with Cultural Sustainability	57
	Becoming a Music Education Researcher in Social Change	58
	Initiating Echoes in the Valley: A Festival for Systems Transformation	60
	Demonstrating Multiple Possibilities Through Public Pedagogy	61
	Festival Practices and Values: Collaboration and Cooperation for Ecological Sustainability	61
	Strategies Towards Sustainable Futures for the Festival	62
	Envisioning Amidst Crises	64
	Looking Around and Looking Forward	64
	A Vision for the Future of Nepali Music	64
	Epilogue	65
	Using the Public Nature of Music-Making for Social Transformation	66
	References	67
6	**Boundary Spanning Ecopolitics**	**69**
	Re-Envisaging Music Education for Ecopolitical Engagements	69
	Beginnings: Listening, Playing, Questioning	69
	An Early Obsession	70
	Career Decisions: Conflicts and Compromise	71
	Being and Becoming an Environmentalist	71
	Becoming a Professional Musician	72
	Shaping a Music Identity	73
	Fusion Musical Identity: Breaking Cultural Barriers	74
	A Moral-Ecological Turn	75
	A Huge Shift: Going Independent	75
	Becoming an Environmental Advocate	76
	Being and Becoming an Advocate	77

	Working With and for Children	77
	Engaging in Public Pedagogy: Affordances and Constraints	79
	Everything is Connected: Systems Transformation for Environmentally Sustainable Futures	82
	Future Challenges	83
	Boundary Spanning in Ecopolitical Professionalism	84
	References	85
7	**Undertaking Systems Transformation Through Ecopolitical Professionalism and Public Pedagogy**	87
	Ecopolitical Systems Thinking in Action	88
	Moving Beyond Path-Dependency for Transformative Music Education Organisations	90
	Lessons in Public Pedagogy for Ecopolitical Professionalism in Music Education	92
	Concluding Remarks	94
	References	95

About the Authors

Margaret S. Barrett is Professor and Head of the Sir Zelman Cowen School of Music and Performance at Monash University and Visiting Professor at the Center of Educational Research and Academic Development (CERADA) of the University of the Arts Helsinki. Her research addresses children's early learning and development in the ecologies of family, community, and school; eco-politics and music education; gender politics in music; narrative and qualitative inquiry methods; pedagogies of creativity, collaboration, and expertise; and singing and invented song-making in children's communities of musical practice. This body of research has been supported by grants from the Australian Research Council, the Australia Council for the Arts, the Australian Children's Music Foundation, the Academy of Finland, the Australian Youth Orchestra, the British Academy, and Musica Viva. She has served as President of the International Society for Music Education, Chair of the World Alliance for Arts Education, and Chair of the Asia-Pacific Symposium for Music Education Research. She has been awarded significant fellowships including a Fulbright Senior Research Fellowship at the Library of Congress and University of Washington, a Fondation de Maison des Sciences de l'Homme Fellowship at the Institute for Research and Coordination in Acoustics/Music (IRCAM), and Beaufort Visiting Research Fellow at St. John's College University of Cambridge. Her publications include the *Oxford Handbook of Early Learning and Development in Music* (2023, with Graham F. Welch), *Collaborative Creative Thought and Practice in Music* (2014), *Narrative Soundings: An Anthology of Narrative Inquiry in Music Education* (2012, with Sandra L. Stauffer), and *A Cultural Psychology of Music Education* (2011).

Margaret Barrett and Heidi Westerlund. © Eeva Anundi

Heidi M. Westerlund is Professor at the Sibelius Academy, University of the Arts Helsinki, Finland, and has also been appointed as Adjunct Professor (Research) at Monash University. Her research interests include higher arts education, music teacher education, collaborative learning, cultural diversity, democracy, and eco-politics and professionalism in music education. She has published widely in international journals and books and is Editor-in-Chief of the Finnish Journal of Music Education. She is Co-editor of *Collaborative Learning in Higher Music Education* (2013, with Helena Gaunt), *Music, Education, and Religion: Intersections and Entanglements* (2019, with Alexis Kallio and Philip Alperson), *Visions for Intercultural Music Teacher Education* (2020, with Sidsel Karlsen and Heidi Partti), *The Politics of Diversity in Music Education* (2021 with Alexis Kallio, Sidsel Karlsen, Kathryn Marsh and Eva Saether), and *Expanding Professionalism in Music and Higher Music Education—A Changing Game* (2021 with Helena Gaunt). She has led several research projects and is currently Lead PI with Margaret Barrett of Music Education, Professionalism, and Eco-Politics (EcoPolitics, 2021–2025) funded by the Academy of Finland.

Abbreviations

CERADA	Center for Educational Research and Academic Development in the Arts
ESD	Education for Sustainable Development
IRCAM	Institute for Research and Coordination in Acoustics/Music
MGIEP	Mahatma Gandhi Institute of Education for Peace and Sustainable Development
NIAS	National Institute of Advanced Studies
NMC	Nepal Music Center
SDG	Sustainable Development Goals
UNCCD	United Nations Convention to Combat Desertification
UNESCO	United Nations Educational, Scientific and Cultural Organization
UNICEF	United Nations Children's Fund
VUCA	Volatile, Uncertain, Complex, and Ambiguous
WEIRD	Western, Educated, Industrial, Rich, Democratic
WOMEX	Worldwide Music Expo

List of Figures

Fig. 4.1	"Hiking the Horizontal" in professional life	52
Fig. 5.1	Public pedagogy and transformative ecopolitical professionalism	67
Fig. 6.1	Boundary spanning between local and global level systems	85

Chapter 1
Introduction

Abstract In an increasingly complex world, confronted by global crises and wicked problems such as climate change, colonialism, geo-political instability, pandemics, poverty and related inequalities, racism, and the rise of political systems that routinely breach human rights conventions, our approach to music education needs rethinking. This chapter outlines a proposal for a systems transformation in music education that shifts the field from an *ego*-logical focus to an *eco*-logical ethos and ecopolitical rationale that acknowledges the professional responsibility of music fields to address the demands of complex concerns of social–ecological and environmental sustainability. The concepts of public pedagogy and professional responsibility are introduced as a means to expand the professionalism of the field. The chapter introduces three narrative accounts, which provide examples of public pedagogy, and concludes with a note on the methodological approach employed in generating these accounts.

> This made me think.
> I am just a drummer. I've always thought that it would be enough.
> But if someone asked me to join in a project that had some specific social or political purpose, I would be happy to join in. Yes, I could do it.
> But I've never thought about this before.

This musician's response to a presentation to an audience of composers concerning the ways in which the music fields can engage with larger societal and global concerns through their practice is illustrative of how music and music education professionals compartmentalise their practice from global societal concerns. The surprise manifested by the drummer (*I've never thought of this before*) and the defence of his position (*I am just a drummer. I've always thought that it would be enough*) captures the ways in which professions may become intellectual straitjackets, governing thought and behaviour and confining us to the discipline and its professional boundaries. In a "silo mentality model" in which "academic boundaries are reinforced by epistemologies and professional boundaries by technical language and practice, the focus is upon the discipline or profession, not the problems with which it has to deal" (Bore & Wright, 2009, pp. 250–251).

We live in an increasingly complex world, confronted by global crises and wicked problems[1] such as climate change, colonialism, geo-political instability, pandemics, poverty and related inequalities, racism, and the rise of political systems that routinely breach human rights conventions. In this world, our approach to music education, including higher music education for musicians and music teachers, needs rethinking. In music education, global policies such as the UNESCO Millennial Goals[2] and UNESCO Sustainable Development Goals[3] are often viewed as political tools for professional advocacy rather than agendas that *call for critical change in the profession itself*. In this book we ask, how do such global concerns intersect with music education theory and practice, and its professional organisations? Why might these concerns be absent? How can music and music education professions proactively influence societal values and processes? These questions have no clear-cut answers, yet they permeate the very core of what (in this book) we call *ecopolitical professionalism in music education*. Our aim is to prompt music educators globally to undertake a paradigmatic shift in perspective, in order to engage with these difficult questions.

As do many educational and arts education researchers (Hunter et al., 2018), we see the need to focus not simply on the "whats," but the "hows" and "whys" of music education—on the *processes* and *values* of music education—and the "where" and "who." We suggest that the volatile, uncertain, complex, and ambiguous (VUCA; van der Steege, 2017) nature of current global contexts, requires the development of *systems thinking in music education* as a key component of both professional and organisational change. Systems thinking (e.g., Checkland, 1981; Checkland & Poulter, 2007; Ison, 2017; Luhmann & Knodt, 1996; Meadows, 2009) provides heuristic tools that equip music educators to think critically in terms of relationships, connectedness, and contexts, to challenge taken-for-granted normative thinking, and to engage with issues beyond the traditional realms of music professions. Crucially, systems thinking assists the music professions to consider their work in relation to wicked problems and to recognise music and music education as socio-political, spatial, material as well as cultural phenomena.

From a systems perspective, in this book we understand music and music education as a large social-ecological system (e.g., Biggs et al., 2022; Folke & Berkes,

---

[1] The etymology of "wicked problems" can be attributed to Rittel and Webber (1973) who were working in the context of urban planning and introduced the term to apply to those problems that are difficult, complex, and resistant to resolution. Whilst science is practised to deal with "tame" problems, wicked problems are those social policy problems that cannot be confronted easily because of the nature of these problems: they are "'malignant' (in contrast to 'benign') or 'vicious' (like a circle) or 'tricky' (like a leprechaun) or 'aggressive' (like a lion, in contrast to the docility of a lamb)"; p. 160). Ritter and Webber present ten defining features of wicked problems: (1) they are resistant to definition; (2) have no end-point; (3) solutions are good-or-bad; (4) there is no immediate or ultimate test of a solution; (5) there is no opportunity to learn by trial and error—solutions are "one-shot operations" and irreversible; (6) there is not an exhaustive set of solutions; (7) they are unique; (8) symptomatic of other problems; (9) explanations of problems determines the nature of the resolution; and (10) for the resolution there is no clear right or wrong (pp. 161–167).

[2] See https://www.who.int/news-room/fact-sheets/detail/millennium-development-goals-(mdgs).

[3] See https://en.unesco.org/sustainabledevelopmentgoals.

1998), composed of a combination of various professional subsystems (e.g., higher music education including teacher education, education of professional musicians, community music, music therapy, and school music education), that also intertwine with other systems such as legal, education, and economical ones. Systems thinking reaches beyond micro-level pedagogical situations in order to assist in recognising the potential blind spots of the profession and its organisations, and the unintended and unwanted consequences that arise from ignoring such blind spots. Systems thinking challenges music and music education professionals to think beyond their individual teaching and learning situations and to consider professional work more broadly as part of larger intersecting systems and in relation to the changing society (see systems thinking in music education, Ilmola-Sheppard et al., 2021; Väkevä et al., 2017, 2022; Westerlund et al., 2021).

Music education has a unique purpose: seen as a social-ecological system, it self-reproduces its boundaries when identifying itself in relation to its environment and other systems (Berger & Luckmann, 1966; Luhmann, 1995). Such reproduction and boundedness is necessary to a degree but can also limit the system's transformative possibilities when the systems environment changes. Systems thinkers argue that in the present complex world all professional systems, including music education, need boundary critique (Ulrich & Reynolds, 2020) to make critical transformation possible. In other words, there is a wider need in society to interrogate established mental models that shape our practices, including their underpinning beliefs, values, and impacts in relation to the changing conditions. By turning a critical lens on the music education professions, we can raise the following questions: What if music education is recognised as part of the problem of continuing unsustainability, inequalities, and even narcissism in society? What if our professional work has unintended and unwanted consequences?

Drawing on these ideas, in this book we seek to prompt systems transformation in our profession—to reconsider the ontological, epistemological, and methodological implications of the systems activities and professional practices—and suggest that we engage in acts of public pedagogy that foster an ecopolitical professionalism in music education and therewith societal transformation. We illustrate the possibilities such a shift offers through presenting three individual narrative accounts of music education change agency and public pedagogy operating at local through to global levels of society in diverse contexts (Chaps. 4, 5, 6). Tuulikki Laes, a social entrepreneur and music education researcher working in Helsinki, Finland, speaks of the ways in which she has "stepped off the path" of music education to use her musical practice to work with individuals and groups who have been marginalised and excluded from music education engagement. We draw on Lerman's notions of *Hiking the Horizontal* (2011) to reflect Laes' boundary-crossing work in music education that seeks to expand understandings of the "who", "how", and "where" of music education participation in Finnish society. Riju Tulahadur, a musician and music teacher in Kathmandu, Nepal describes the ways in which he works in multiple positions with communities to address humanitarian goals, effect social transformation, and foster gender-inclusive cultural sustainability, while working to improve the professional employment possibilities for Nepali musicians. Ricky Kej, a composer, environmentalist, and multiple Grammy awardee based in Bengaluru, India describes how he uses his music-making to advocate for change in global political forums. These individuals' accounts of their personal stories, professional values, struggles, and achievements illustrate the diversity of positionings of professional musicians and music educators who work across intersecting social–ecological systems, and cultural and political contexts around the globe. We trace these pathways through musicianship, music education, and active change agency from their early beginnings, not only to illustrate this diversity but also to demonstrate the interconnected, whole-of-life nature of emerging active change agency.

Through the presentation and analysis of the three narrative accounts (Barrett & Stauffer, 2009, 2012) we challenge an "imperialist professionalism" that assumes all music educators have "the onus to conform to the academic practices privileged in Western institutions" (Tran & Nguyen, 2015, p. 962). As Sidsel Karlsen (2019) put it:

> there are numerous problems connected to scholars speaking for others, but … this should not prevent us from using our professional powers to allow others to have their voices heard. As teachers, editors, reviewers, examiners and opponents, we have the opportunity to allow or deny expressions of vulnerability to exist within the discourse of music-education practice, scholarship and research, both on the individual and the systemic level. While such expressions should of course never be the source of a positive evaluation alone, we definitely have a choice regarding whether or not we consider them as valuable or legitimate contributions. (p. 192)

In this way we present music education professionals as working in particular places and particular times, responding to the specific needs of individuals and communities, bringing and bearing a range of moral and ethical responsibilities in their professional work.

Professional Responsibility as an Imaginative Possibility

We aim to *expand music education's professional horizons* to embrace *a new ethos and rationale* that acknowledges the professional responsibility of the music fields to address the demands of complex concerns of social–ecological and environmental sustainability. Professional responsibility has been viewed as a form of moral responsibility in the sense of an obligation and decision-making that calls forth "a sense of duty to care for self and others, extending beyond one's own self-interest, and accountability to others for one's actions" (Fenwick, 2016, p. 4). While professionalism as a concept centres on competencies and work in occupations, an ethical and moral responsibility is inherent, relating professional work with "service" towards a community or state (Carr, 2014; Minnameier, 2014). Such responsibility asks all professions to navigate between the "near and far," the horizontal issues, to engage in a morally informed professional praxis that recognises individual and collective responsibilities in working towards resolution of global wicked problems. In this way, *professional practice as praxis* in its critical mode (Westerlund & Partti, 2018) is responding to various site-based needs and interests in a way that "adds a moral and social justice dimension to the enactment of educational praxis" (Mahon et al., 2020, p. 23). A shift from practice to critical praxis which is underpinned by systems understanding, thus repositions music and music education professionalism, and highlights our ethical and moral professional responsibilities beyond a narrow focus on issues of musical quality, genre-related musical criteria and traditions.

Historically, the term profession has been used to distinguish between occupations that require academic training and adherence to a codified set of standards (e.g., law, accountancy, medicine) and those that do not, including teaching and the arts (Cribb & Gewirtz, 2015). In the literature, professionalism is typically used to refer to "the competence and expertise of an individual and the quality of the work they do" (Alexander et al., 2019, pp. 2, 3). In this way, professionalism becomes the central concept in building upon the foundational constitutive relationship between the whole of society and an occupational group (e.g., Cribb & Gewirtz, 2015; Fenwick, 2016; Minnameier, 2014; Solbrekke & Sugrue, 2011). To claim "professionalism" in music education thus means that the conduct, aims, values, responsibilities, and ongoing development of practising professionals is conceived in relation to the whole of society and its challenges and policies, institutional settings, and horizontal changes (Westerlund & Gaunt, 2021; Westerlund et al., 2021). In this view, various societal concerns are seen as part of professionalism instead of some extra added aspect that can equally well be put aside. While we recognise the emphasis on musical expertise in professional education as a legitimate element of music and music education professionalism, as argued by the sociologists of professionalism (e.g., Abbott, 1988), we will at the same time highlight how individuals and groups with no formal professional education in music may pave the way for understanding the wider potential of music and the music education professions in transforming societies.

We build on these general understandings of professionalism to propose a view of music and music education professionalism as future-oriented, emancipatory, democratic, collaborative, and committed to relational renewal and ecological responsibility. This proposal is underpinned by the following questions: *How might current and historical music and music education practices be complicit in maintaining unsustainability and inequalities in society? What might music teaching, teacher education, and professional work in music look like when developed through dialogue with an ecological futures perspective? What might be the outcomes of nurturing and celebrating the public nature of music in human life?* In posing these questions we seek to provoke further dialogue rather than arrive at singular definitive responses or models.

Music Education, Ecopolitical professionalism and Public Pedagogy: Towards Systems Transformation thus suggests a political and moral shift that recognises music education not just as a cultural endeavour, but also as a public, social, and spatial endeavour that exercises active agency in society. We address the need for the music education professions to interrogate the spaces where education takes place as well as the practices and professional ways of thinking to envision alternative "worlds," unorthodox "realities" and shifted "regimes of truth" (Foucault, 1978). By striving for boundary critique and systems reflexivity (Westerlund et al., 2021) in how music education operates, specifically through the reciprocal interrogation of its values and the ways these shape society, we suggest that it is possible to develop the capacity to transform the profession's relationship to society with a sense of responsibility. The book thus promotes a proactive professional stance, in which "experimenting" with new ideas and creating new spaces becomes an inherent rule of "game-changing" rather than a rare exception (Gaunt & Westerlund, 2021).

This kind of stance as part of professional work asks music educators to consider the virtues of a profession and the political values that might resonate with professional responsibilities that range beyond the boundaries of existing practices and known traditions. As work in other professional fields is increasingly becoming entangled with various societal systems, such as the economy or larger education systems, professionals in music education are "confronted with aspects of a polycentric society that has more than one rationality, logic or locus of reflection" creating tensions, insecurities, and conflicting expectations (Vogd, 2017, n.p.). In this relational "parliament of things" no field can simply apply "value-free, technically defined authoritatively prescribed competences" (Cribb & Gewirtz, 2015, p. 73). The work of music educators is therefore seen as part of emergent and translational processes and becomings through which professionals enact visions and imaginary realities.

Such "expanding professionalism" (Westerlund & Gaunt, 2021) thus reaches beyond what Donald Schön (1983/2016) called mere "technical rationality" to position music educators as systems thinkers and change agents. As change agents, musicians and music educators are encouraged to engage with societal concerns rather than simply offer what the (e.g., economically oriented) society seems to want and what the established ideas of the field demand. Crucially, they can resist, be "obstinate," using Gert Biesta's (2019) terms, in creative and imaginative ways. A mature profession develops thus a multidimensional transformative view of professionalism in which the environment is transformed as well as the professional work itself.

A Shift from the Ego-Logical to the Eco-Logical

In pursuing the ideas outlined above, we apply an ecological perspective to music education professionalism in order to shift the ego-logical and ego-centric discourse of the musical self to an ecological discourse of the world and its interrelated and interconnected ecosystems. In searching for an *eco*-logical shift to reveal and remedy the current situation, we suggest expanding the professional horizons concerning who music educators are in society and how they engage in their actions in the world. We understand the environment in which music teachers and music teacher educators work not simply as surrounding music education, but as a zone of interpenetration in which lives are comprehensively entangled (e.g., Ingold, 2011).

By using statements regarding environmental concerns (including the dramatic consequences of the COVID-19 pandemic and policy concerns of UNESCO's Education for Sustainable Development (ESD; UNESCO, 2014)) as a stepping stone, we consider social–eco-logical concerns, thereby expanding ecopolitical ecology and its implications into multiple social arenas and horizons to enable societal transformation towards environmental sustainability, nonviolence, equity, social justice, and grassroots democracy. Such eco-logically oriented politics in music education professionalism aims to engage music educators with "near" and "far" change and deals with complex everyday processes, values, and ways of thinking, being, and doing. Through ecopolitical processes, music teachers exercise agency, negotiate power, and use imagination responsibly as integral parts of the intersubjective intertwining ecosystems in which they work. Such ecopolitics in music education thus deals not simply with musical challenges and technical issues but also with wicked problems (Rittel & Webber, 1973)—such as climate change and the effects of poverty—that are not just the responsibility of state level politicians. Rather, we argue that individual professionals and organisations and the entire music education system need to expand professional horizons and experience to engage with wicked problems; they need to collaborate and build shared understandings among relevant stakeholders and organisations as well as "continuously explore boundaries among the surrounding systems to understand the values, processes and behaviours relevant to the problems, and how they change over time" (Gonzales, 2020, p. 28).[4]

By and large ecological thinking and ecopolitics might be viewed as a hybrid concept, in which one foot is "in the realm of the 'real', and one foot in the realm of ideas; and the two realms play against each other" (Barnett, 2017, p. 20). The ecological perspective:

> may be attributed in part to a desire to move beyond Western individualist explanations of human thought and activity to recognition of more diverse and pluralist accounts. Such a move implies also the desire to move beyond the 'mechanistic, reductionist, and atomistic' (Capra, 1996) forms of investigation that have tended to dominate science to embrace more holistic and divergent forms of investigation. (Barrett, 2012, p. 207)

In this way, an ecopolitical systems approach to music education distances itself from behaviourist psychology and understanding of knowledge in which education appears as neutral and epistemologically objective. As Margaret Barrett (2012) wrote, ecological thinking interrogates "the ways in which modernist Western epistemologies of individualism and mastery legitimate the subjugation and exploitation of peoples" in order to better understand how "place and location" shape, celebrate, and support diversity (p. 208).

The book combines ecological systems thinking with professionalism based on an observation that ecological thinking beyond this book's foci and limited space may have much to contribute to our critical understanding of:

- questions of sustainability and how the relationships between humans and nature relate to music;

---

[4] This kind of boundary-crossing expansion has previously been suggested by several scholars in education (e.g., Akkerman & Backer, 2011; Engeström, 1987; Gonzales, 2020; Sachs, 2003) as well as music education (e.g., Koivisto, 2022; Lehtinen-Schnabel, 2022; Miettinen, 2019; Westerlund et al., 2019).

- professional responsibilities of music educators to consider the ways in which broad social and cultural factors shape and can be shaped by the physical and biological environment, and the human relationships and interactions that occur within these responsibilities;
- music engagement and learning being and being acted upon by the ecosystems within which they are situated;
- the ways ecological injustices might permeate musical practices and music education institutions and organisations as social endeavours, including how patriarchal structures continue to exclude women and girls;
- how settler colonial structures systematically disenfranchise Indigenous peoples;
- how the goals of sustainability and music education professionalism can be intertwined by highlighting the value of the work conducted in the Global South and the majority world;
- the policy–practice relationship being about multilevel transformation;
- the potential of music and music education being a public endeavour and a form of public pedagogy not just within individual and collective–communal ecologies but also macrolevel political arenas;
- active efforts by music and music education professionals towards promoting ecological justice worldwide.

Our approach to an ecopolitical music education professionalism moves beyond ecological concerns that are limited to questions of the sustainability of endangered musical heritage (Schippers & Grant, 2016) to acknowledge our responsibilities to the intersecting ecologies of the natural, social, and political worlds in which we live. Our ecopolitical, ecosocial shift resonates with the emerging discourse of artistic citizenship (Elliott et al., 2016), place-making in education (Stauffer, 2009, 2012), eco-literate music pedagogy (Shevock, 2018), social justice and activism (Hess, 2019), as well as the arguments proposed in EcoJustice Education (Martusewicz, 2019) and "sustainable arts" in education (e.g., environmental arts education, or environmental education as artistic practice) movements. We address these questions as *questions of professional responsibility in music education* located in complex systems: music educators are seen as not merely reflecting the world but acting to transform it while *transforming their own professional practice and the profession itself*.

The Potential of Public Pedagogy

As a whole, this volume combines an expanding ecological understanding of professionalism with the potential for public pedagogy as a key element of ecopolitical professional work in music education. The concept of public pedagogy might be viewed as essentially contested. As Michael O'Malley and colleagues wrote:

> There is no consensus on how to define public pedagogy, but educational scholars employing the term typically explicate its feminist, critical, cultural, performative, and/or activist pedagogical dimensions. Other scholars take up the challenge of redefining education in order to deinstitutionalize its conceptualization and uncouple it from its automatic associations with schools; and yet others take these criticisms further to explore posthuman reconceptualizations of pedagogy. (O'Malley et al., 2020, p. 1)

We recognise the public nature of music and the ways in which public performances create images in society concerning for whom music education is meant, how one is expected to engage with it, and what its purposes are. Put simply, every music performance is potentially a form of public pedagogy whether we are aware of the implicit meanings about what music is, who it is for, where it might be enacted, and why. Considering music performance and presentation in this way alerts us to the potentially negative messages that might be drawn from that experience. By "public pedagogy" we refer to music educators' work that makes space for interrelations and resonances between politics, arts, and education, and reopens the public sphere for participants "beyond pedagogy as a cognitive and rational process of transmission" (Sandlin et al., 2017, p. 827). Writing in 2012, music educators Randall Allsup and Eric Shieh called for a "public music pedagogy" that recognised teachers' "moral imperative to care" (p. 48) and underpinned a social justice approach to music education. They advocated for a music education that embraces "going public, coming out" (p. 50) in order to address the "big questions of our time" (p. 51). Subsequent work in areas such as music education activism (Hess, 2019), decolonisation of the music curriculum (Philpott, 2022), embedding Indigenous music in schools (Prest et al., 2022) and expanding professionalism through participatory music-making in new places and spaces (Westerlund & Gaunt, 2021) might be seen as responses to this call.

We reach beyond the dominant discourses of public pedagogy that have focused on the media, popular culture, and society as "educative forces," situating music as an instructive medium for the transmission of key ideas and messages. Rather, we consider the various ways music as public art can contribute to "social and cultural geographies of place and identity" (Schuermans et al., 2012, p. 5), individual and collective agency in the public world, as well as systems transformation of institutionalised forms of music education. In this way public pedagogy advocates transformative action on multiple levels and in multiple locations; from levels of individual interests, emotions, and ethical commitments to those of broader socio-political and societal issues, in and beyond the institutions of schools and schooling.

We draw on the work of cultural geographers Nick Schuermans and colleagues to explore the possibilities of the ways in which public pedagogy can redirect the focus from the meaning "embedded" in musical works (including the technical questions of right and wrong) towards considering the sociopolitical processes stimulated by musical performance, practice, and learning. Public pedagogy might then focus on the ways music "opens up new ways of seeing, feeling, experiencing, and describing the world" (Schuermans et al., 2012, p. 677) thereby expanding societal horizons and music educators' understandings of their profession's contributions to society. The three narratives presented in this book illustrate how musicians can aim to undertake "deliberative, active political interventions in society," emphasise "that there is 'educational work' to be done in and for the public sphere," and focus "on the concrete practices of citizens engaged corporeally in social interactions which unsettle established notions of living together" (p. 677). As Schuermans and colleagues argued:

> public pedagogy provides us with the theoretical tools to disentangle the plural, shifting, open, and contested ways of the socio-political "making of places" by artists and to understand their role as "educators" when intervening in localised struggles for more freedom, more equality, or better citizenship rights. (p. 679)

Public pedagogy as integrated in an ecological understanding of professional work can thus provide us with "insights to conceive of and strive for social and spatial justice" (p. 680). Public pedagogy can be seen as a powerful way to reconceptualise and redefine education, both inside and outside institutions in order to engage children, music teachers, and their communities in transformative change. The potential of music education in public pedagogy practice, we suggest, lies in its performative nature and location in public and private spaces and places. Through this book we ask: How might this potential be better realised?

A Note on the Research Approach

In order to illustrate ecopolitical systems thinking and public pedagogy in music and music education we adopted a narrative inquiry approach to generate the three narratives that are featured in Chaps. 4, 5, 6. Narrative inquiry is underpinned by a relational ontology and is described as a relational methodology for studying experience (Clandinin, 2022). Narrative inquiry therefore provides a co-creative process of data generation, analysis and storied presentation in which researcher and research participants work in relationship through the inquiry process. Narrative inquiry generates stories of experience which are subsequently re-storied collaboratively, thereby providing "…not only... a privileged vehicle for exploring the human realm, (but also) a primary inroad into understanding human life and human selfhood" (Freeman, 2012, p. 25). Interviews, particularly life-history interviews, are a primary method for data generation in narrative inquiry, supported by other qualitative methods such as observation and artefact analysis.

In this inquiry we aimed to interrogate the ecopolitical and public pedagogy beliefs, values and practices of three music education practitioners and to understand their work as transformative systems practitioners. We had been familiar with the work of the three participants, Tuulikki Laes, Riju Tulahadur, and Ricky Kej over a period ranging from 7 (Kej) to fifteen years (Laes). Our invitations to them to participate in the research were therefore based on long-standing near and far observation of their practice. For example Heidi has worked with both Tuulikki and Riju in a range of projects. Margaret has been familiar with Ricky's work for some years, extending invitations for him to speak on this work in a range of forums including the first ISME South Asia Regional Conference held in Bangalore in 2017. We are indebted to each of them for their willing participation in the research, their generosity in reviewing interview transcripts and multiple drafts of the narratives, and endorsement of the final texts.

Extended life history interviews were conducted with each participant via zoom in late 2020 and the first months of 2021. Interviews with Riju and Ricky were conducted as three way conversations led by Margaret and with both researchers participating. Margaret undertook the interview with Tuulikki alone given Heidi's long-standing relationship as Masters and Doctoral supervisor, friend, and now colleague. Each participant reviewed the transcripts of the interviews to confirm these as an accurate record of our conversations. Margaret subsequently undertook a narrative analysis of the transcripts to identify key events and emerging themes in relation to ecopolitical and public pedagogy beliefs, values and practices and produce a storied account of these themes through a life-history lens. Heidi checked these first narrative accounts with both researchers then working together to refine the texts prior to sending the relevant draft to each participant. Tuulikki, Riju and Ricky reviewed the first drafts providing feedback, clarifying details, and making small changes (e.g., naming of places, changes of dates). Margaret and Heidi then undertook a further analysis of each narrative account in order to identify and make explicit the intersecting social-ecological systems and the processes of systems transformation in each narrative. The final version of each narrative presented in this text has been checked and approved by each participant.

The narratives draw on the principles of 'resonant work' in narrative inquiry (Barrett & Stauffer, 2009, 2012); those of respect, responsibility, resilience, and rigour. We have endeavoured to ensure that our work with Tuulikki, Riju and Ricky has been respectful through maintaining a relational and consultative focus throughout the research process. We have been alert to our responsibilities not only to these participants, but also to the music education professions in the ways that we have shaped their narrative accounts. In presenting these narratives we hope they will be resilient, in the sense of being accounts that will endure and continue to provoke and inform the profession. Finally, in outlining our processes in detail we seek to demonstrate rigour in how we have undertaken this research.[5]

References

Abbott, A. (1988). *The system of professions: An essay on the division of expert labor*. University of Chicago Press.

Akkerman, S. F., & Bakker, A. (2011). Boundary crossing and boundary objects. *Review of Educational Research, 81*, 132–169. https://doi.org/10.3102/0034654311404435.

---

[5] Ethical permission to undertake the research was granted by the Human Research Ethics Committee at Monash University (Application No 26669 approved October 26, 2020).

Alexander, C., Fox, J., & Gutierrez, A. (2019). Conceptualising teacher professionalism. In A. Gutierrez, J. Fox, & C. Alexander (Eds.), *Professionalism and teacher education: Voices from policy and practice* (pp. 1–23). Springer Nature.

Allsup, R., & Shieh, E. (2012). Social justice and music education: The call for a public pedagogy. *Music Educators Journal, 98*(47), 47–51.

Barnett, R. (2017). *The ecological university: A feasible utopia* (1st ed.). Routledge.

Barrett, M. S., & Stauffer, S. L. (Eds.). (2009). *Narrative inquiry in music education: Troubling certainty*. Springer.

Barrett, M. S., & Stauffer, S. L. (Eds.). (2012). *Narrative soundings: An anthology of narrative inquiry in music education*. Springer.

Barrett, M. S. (2012). Troubling the creative imaginary: Some possibilities of ecological thinking for music and learning. In D. J. Hargreaves, D. Miell, & R. A. R. MacDonald (Eds.), *Musical imaginations: Multidisciplinary perspectives on creativity, performance, and perception* (pp. 206–219). Oxford University Press.

Berger, P. L. & Luckmann, T. (1966). *The social construction of reality: A treatise in the sociology of knowledge*. Doubleday Anchor Books.

Biesta, G. (2019). *Obstinate education: Reconnecting school and society*. Leiden.

Biggs, R., de Vos, A., Preiser, R., Clements, H., Maciejewski, K., & Schlüter, H. (Eds.). (2022). *The Routledge Handbook of Research Methods for Social-Ecological Systems*. Routledge.

Bore, A., & Wright, N. (2009). The wicked and complex in education: Developing a transdisciplinary perspective for policy formulation, implementation and professional practice. *Journal of Education for Teaching, 35*(3), 241–256.

Capra, F. (1996). *The web of life: A new scientific understanding of living systems*. New HarperCollins.

Carr, D. (2014). Professionalism, profession and professional conduct: Towards a basic logical and ethical geography. In S. Billett, C. Harteis, & H. Gruber (Eds.), *International handbook of research in professional and practice-based learning* (pp. 527). Springer.

Checkland, P. B. (1981). *Systems thinking, systems practice*. Wiley.

Checkland, P., & Poulter, J. (2007). *Learning for action: A short definitive account of soft systems methodology, and its use for practitioners, teachers and students*. Wiley.

Clandinin, D. J. (2022). *Engaging in narrative inquiry (2nd Edition)*. Routledge.

Cribb, A., & Gewirtz, S. (2015). *Professionalism*. Polity.

Elliott, D., Silverman, M., & Bowman, W. (Eds.). (2016). *Artistic citizenship: Artistry, social responsibility, and ethical praxis*. Oxford University Press.

Engeström, Y. (1987). *Learning by expanding: An activity theoretical approach to developmental research*. Orienta-Konsultit.

Fenwick, T. (2016). *Professional responsibility and professionalism: A sociomaterial examination*. Routledge.

Folke, C., & Berkes, F. (1998). *Understanding dynamics of ecosystem-institution linkages for building resilience*. Beijer Discussion Paper No. 112. Beijer Institute.

Foucault, M. (1978). *Discipline and punish: The birth of the prison*. Pantheon.

Freeman, M. (2012). *Hindsight: The promise and peril of looking backward*. Oxford University Press.

Gaunt, H., & Westerlund, H. (2021). Invitation. In H. Westerlund & H. Gaunt (Eds.), *Expanding professionalism in music, performing arts, and higher arts education: A changing game* (pp. xiii–xxxiii). Routledge. https://doi.org/10.4324/9781003108337.

Gonzales, M. (2020). *Systems thinking for supporting students with special needs and disabilities: A handbook for classroom teachers*. Springer.

Hess, J. (2019). *Music education for social change: Constructing an activist music education*. Routledge.

Hunter, M. A., Aprill, A., Hill, A., & Emery, S. (2018). *Education, arts and sustainability: Emerging practice for a changing world*. Springer Briefs for Education.

References

Ilmola-Sheppard, L., Rautiainen, P., Westerlund, H., Lehikoinen, K., Karttunen, S., Juntunen, M.-L. & Anttila, E. (2021). *ArtsEqual: Equality as the future path for the arts and arts education services.* https://urn.fi/URN:ISBN:978-952-353-043-0.

Ingold, T. (2011). *The perception of the environment. Essays on livelihood, dwelling and skill.* Routledge.

Ison, R. (2017). *Systems practice: How to act in situations of uncertainty and complexity in a climate-change world* (2nd ed.). Springer.

Karlsen, S. (2019). Competency nomads, resilience and agency: Music education (activism) in a time of neoliberalism. *Music Education Research, 21*(2), 185–196. https://doi.org/10.1080/14613808.2018.1564900.

Koivisto, T. (2022). *The (un)settled space of healthcare musicians: Hybrid music professionalism in the Finnish healthcare system* [Doctoral dissertation]. Studia musica 89. Sibelius Academy. University of the Arts Helsinki. https://urn.fi/URN:ISBN:978-952-329-264-2.

Lerman, L. (2011). *Hiking the horizontal: Field notes from a choreographer.* Wesleyan University Press.

Luhmann, N. (1995). *Social systems.* Stanford University Press.

Luhmann, N., & Knodt, E. M. (1996). *Social Systems.* Stanford University Press.

Mahon, K., Edwards-Groves, C., Francisco, S., Kaukko, M., Kemmis, S., & Petrie, K. (2020). *Pedagogy, education, and praxis in critical times.* Springer Nature.

Martusewicz, R. A. (2019). *A pedagogy of responsibility.* Routledge.

Meadows, D. H. (2009). *Thinking in systems: a primer.* Earthscan.

Miettinen, L. (2019). Religious identities intersecting higher music education. An Israeli music teacher educator as a boundary worker. In A. Kallio, P. Alperson, & H. Westerlund (Eds.), *Music, education, and religion: Intersections and entanglements* (pp. 233–248). Indiana University Press.

Minnameier, G. (2014). Moral aspects of professions and professional practice. In S. Billett, C. Harteis, & H. Gruber (Eds.), *International handbook of research in professional and practice-based learning* (pp. 57–77). Springer.

O'Malley, M. P., Sandlin, J. A., & Burdick, J. (2020). Public pedagogy theories, methodologies, and ethics. *Oxford research encyclopedia of education* (pp. 1–25). https://doi.org/10.1093/acrefore/9780190264093.013.1131.

Philpott, C. (2022). What does it mean to decolonise the school music curriculum? *London Review of Education, 20*(1). https://doi.org/10.14324/LRE.20.1.07.

Prest, A., Goble, S., & Vasquez-Cordoba, H. (2022). On embedding Indigenous musics in schools: Examining the applicability of possible models to one school district's approach. *Update: Applications of Research in Music Education, 41*(2), 60–69. https://doi.org/10.1177/87551233221085739.

Rittel, H. W. J., & Webber, M. M. (1973). Dilemmas in a general theory of planning. *Policy Sciences, 4*, 155–169.

Sachs, J. (2003). *The activist teaching profession.* Open University Press.

Sandlin, J. A., Burdick, J., & Rich, E. (2017). Problematizing public engagement within public pedagogy research and practice. *Discourse: Studies in the Cultural Politics of Education, 38*(6), 823–835.

Schippers, H., & Grant, C. (2016). *Sustainable futures for music cultures: An ecological perspective.* Oxford University Press.

Schön, D. (1983/2016). *The reflective practitioner: How professionals think in action.* Routledge.

Schuermans, N., Loopmans, M., & Vandenabeele, J. (2012). Public space, public art and public pedagogy. *Social and Cultural Geography, 13*(7), 675–682.

Shevock, D. J. (2018). *Eco-literate music pedagogy.* Routledge.

Solbrekke, T. D., & Sugrue, C. (2011). Professional responsibility—back to the future. In C. Sugrue & T. D. Solbrekke (Eds.), *Professional responsibility. New horizons of praxis* (pp. 11–28). Routledge.

Stauffer, S. L. (2009). Placing curriculum in music. In T. Regelski & J. T. Gates (Eds.), *Music education for changing times: Guiding visions for practice* (pp. 175–186). Springer.

Stauffer, S. L. (2012). Place, music education, and the practice and pedagogy of philosophy. In W. D. Bowman & A. L. Frega (Eds.), *The Oxford handbook of philosophy in music education* (pp. 434–452). Oxford University Press.

Tran, L. T., & Nguyen, N. T. (2015). Re-imagining teachers' identity and professionalism under the condition of international education. *Teachers and Teaching: Theory and Practice, 21*(8), 958–973. https://doi.org/10.1080/13540602.2015.1005866.

Ulrich, W., & Reynolds, M. (2020). Critical systems heuristics: The idea and practice of boundary critique. In M. Reynolds & S. Holwell (Eds.), *Systems approaches to making change: A practical guide* (pp. 255–306). Springer.

UNESCO. (2014). *UNESCO roadmap for implementing the global action programme on education for sustainable development*. https://unesdoc.unesco.org/ark:/48223/pf0000230514.

Väkevä, L., Westerlund, H., & Ilmola-Sheppard, L. (2017). Social innovations in music education: Creating institutional resilience for increasing social justice. *Action, Criticism, and Theory for Music Education, 16*(3), 129–147. https://doi.org/10.22176/act16.3.129.

Väkevä, L., Westerlund, H., & Ilmola-Sheppard, L. (2022). Hidden elitism: The meritocratic discourse of free choice in Finnish music education system. *Music Education Research, 24*(4), 417–429. https://doi.org/10.1080/14613808.2022.2074384.

Vogd, W. (2017). The professions in modernity and the society of the future: A theoretical approach to understanding the polyvalent logics of professional work. *Professions & Professionalism, 7*(1), Article e1611. https://doi.org/10.7577/pp.1611.

Westerlund, H., & Gaunt, H. (Eds.). (2021). *Expanding professionalism in music and higher music education: A changing game* (1st ed.). Routledge. https://doi.org/10.4324/9781003108337.

Westerlund, H., Väkevä, L., & Ilmola-Sheppard, L. (2019). How music schools justify themselves: Meeting the Social challenges of the 21st century. In M. Hahn & F.-O. Hofecker (Eds.), *The Future of Music School–European Perspectives* (pp. 15–33). Musikschul management Niederösterreich GmbH. https://doi.org/10.21939/future_of_music_schools.

Westerlund, H., Karttunen, S., Lehikoinen, K., Laes, T., Väkevä, L., & Anttila, E. (2021). Expanding professional responsibility in arts education: Social innovations paving the way for systems reflexivity. *International Journal of Education and the Arts, 22*(8). https://doi.org/10.26209/ijea22n8.

Westerlund, H., & Partti, H. (2018). A cosmopolitan culture-bearer as activist: Striving for gender inclusion in Nepali music education. *International Journal of Music Education, 36*(4), 531–546. https://doi.org/10.1177/0255761418771094.

Chapter 2
The Emerging Ecological Shift

Abstract This chapter explores the emerging ecological shift in music education theory and practice. The chapter presents a critical view of ecological thinking and the need to recognise and address the unintended consequences of cultural and educational practices. The chapter examines global sustainability agendas such as the United Nations Sustainable Development Goals in order to move beyond inward looking perspectives of sustainability as conservation of existing practices and to re-frame music education. The chapter calls for an ecological awareness which recognises that social-ecological "wicked problems" are not simply issues of our ecological footprint; in short, that artists are ecological whether they are aware of this or not.

> Reframing the ecological crisis at smaller scales so as to question our personal relationship with our fellow human beings and with non-humans is a great occasion to reconsider the ways we can accommodate and welcome the otherness of others. At the global scale, the human imprint on the Earth might seem omnipresent and unprecedented, but at a smaller scale one just needs to let things go in a couple of square metres in the backyard to realise that nature is neither dead nor agonising. Nature is everywhere, latent, silent, waiting for the opportunity to burgeon and flourish, and there are many ways to protect it other than to subjugate it. (Maris, 2015, pp. 131–132)

In this chapter we begin with Virginie Maris' (2015) call to reframe the ecological crisis at a smaller scale. In her work she identifies three features of the Anthropocene sciences, those of "their global scale, humanity as a species, and the techno-scientific characterisation of the problems" as complicit in dispossessing individuals "of their moral responsibility and ability to be actors in the solutions" (p. 130). The sheer scale of these features can render us helpless in considering ways in which we can engage in our professional work through an ecological lens. In this chapter we explore understandings of ecology and their application in music education in order to consider the implications for professionalism theory and practice in music and music education.

Working "Across Scales"

The first documented use of the term *ecology* was by German zoologist Ernst Haeckel in 1866 to distinguish between "the inclusive study of organisms in the environment, in contradistinction to the narrower study of organisms in the laboratory" (Keller & Golley, 2000, p. 9). The gradual development of ecology as a field of study has seen a move from the distinct study of animal ecology and the natural world to encompass human ecology and consideration of the ways in which these intersect (see, e.g., F. Fraser Darling's, 1951 proposal of an Ecological Approach to the Social Sciences). Ecological definitions and descriptions focus on the *relationality* between organisms and the environments in which they live. Within developmental psychology for example, Bronfenbrenner's ecological systems theory demonstrates the intersecting social environments that shape children's development (1974, 1977).

The need for ecological thinking is not new. Some hundred years ago, educational philosopher John Dewey (1938/1986) argued human beings and their actions are integral parts of their environment:

> There is, of course, a natural world that exists independently of the organism, but this world is environment only as it enters directly and indirectly into life-functions. The organism is itself a part of the larger natural world and exists as organism only in active connections with its environment. Integration is more fundamental than is the distinction designated by interaction of organism and environment. The latter is indicative of a partial disintegration of a prior integration, but one which is of such a dynamic nature that it moves (as long as life continues) toward reintegration. (p. 40)

In short, an ecological understanding of the world points to the interconnectedness of all things in the world. In this sense one could argue, as has philosopher Timothy Morton (2018), that all music and art is in fact ecological because art "includes its environment(s) in its very form" (p. 18). Such inclusion of the environment does not however ensure that art recognises the ecological aspects of its own existence, or aims at ecological ethics. As Morton argued, not all art is ecologically aware, since "ecological awareness is awareness of unintended consequences" (p. 16). He has concluded that:

> the problem with ecological awareness and action isn't that it's horribly difficult. It's that it's too easy. … You are breathing air, your bacterial microbiome is humming away, evolution is silently unfolding in the background. Somewhere, a bird is singing and clouds pass over-head. … You don't have to *be* ecological. Because you *are* ecological. (p. 105)

The problem arises when artists do not recognise that they *are* ecological, part of an ecospheric whole; instead, by focusing only on themselves a form of "ecological blindness" (Plumwood, 2001) takes hold. In music education this might be manifested in a pedagogical approach that arises from a singular universalist perspective, and which is applied regardless of the people, places, and communities in which this education takes place. An ecological awareness recognises that our environment is also social and that social-ecological problems are not simply issues of our ecological footprint but are "inextricably woven through issues of justice and fairness" (Dyball & Newell, 2015, p. 11). Ecological thinking needs to be holistic in the sense that sustainability cannot be demystified unless we examine it together with our worldviews, ethics, and values (Washington, 2015, p. 136). "'Ecology' is fact and values together," as higher education philosopher Ronald Barnett (2017, p. 17) put it. Ecological thinking and acting can therefore not be reduced to singular matters such as offsetting our CO_2 emissions, although finding solutions for our ongoing environmental abuse is central and of utmost importance. As Morton (2018) has reminded us, "Ecological awareness means thinking and acting ethically and politically on a lot of scales, not just one" (pp. 32–33).

Ecological Thinking Beyond Sustainability

Linked with the concept of ecological thinking and acting is the notion of sustainability; how the ecological systems within which we live continue to flourish and maintain diversity. The Sustainable Development Goals (SDGs) were established by the United Nations in 2015 as a "call to action to end poverty, protect the planet, and ensure that by 2030 all people enjoy peace and prosperity" (United Nations Development Program [UNDP], 2022). Key to achieving these goals is recognition of the ways in which action in one area has ramifications for others, that "development must balance social, economic and environmental sustainability" (UNDP, 2022). A critique of the SDGs (Feeney, 2020) identified a number of tensions and inherent problems related to scope, ambition, accountability, inequities, trade-offs, costs, and monitoring. These include the critical trade-offs that need to be made between a goal to end poverty and inequality, and the inevitable environmental impacts of "undermining the conservation of land, efforts at reducing carbon emissions and initiatives to improve access to safe and affordable water" (pp. 343–351) in order to feed the world. One approach recommends tailoring such goals to address local and national needs rather than a universal application (Feeney, 2020). Whilst this might be a solution, such an approach has inherent problems: the capacities for nations to pick and choose which goals to address, and how, can intensify the inequities between the "majority world" and WEIRD (Western, Educated, Industrial, Rich, Democratic) societies.

The debates surrounding the SDGs illustrate the ways in which sustainability is a contested concept despite the central role it has played in establishing environmentalism (Blüdorn & Welsh, 2007). As researchers of social sustainability Ingolfur Blüdorn and Ian Welsh (2007) have noted:

> It is now a commonplace to distinguish between different forms of sustainable development and sustainability ... [as] [t]he prime distinction between "strong" and "weak" forms involves differences in emphasis placed on inter-generational equity, North–South equity and the importance attached to precaution within regulatory and legislative institutions. (p. 189)

Writing some 17 years ago, these authors cautioned that as individuals perceive an escalating burden of "self-responsibility" "environmental issues are delegated to political actors and regulatory regimes" (p. 189) rather than resting at the individual level. In this presentation of sustainability and sustainable development as occurring in "strong" and "weak" forms leading to different emphases on issues of equity, we are again reminded of the need to think and act across multiple scales of activity and of the dangers of the development of a "politics of unsustainability." In his later writings on the exhaustion of the sustainability paradigm, manifested in the "politics of unsustainability," Blüdorn (2017) described these as "discourses of simulation ... [that] help to organise—quite contrary to their own self-perception and declared intentions—modern societies' journey towards ever more social inequality and ecological destruction" (p. 43).

As did Morton, Barnett (2017) has highlighted the *critical* side of ecological thinking and the need to recognise and address the unintended consequences of cultural and educational practices. Barnett wrote:

> Ecology is about the sense of systems—ecosystems—that have been corrupted in some way; their inner diversity has been weakened or some other misfortune has befallen them. It also contains an inner commentary that humanity bears some responsibility for such disturbances and an urging that humanity has a responsibility to help in restoring the integrity of its environment. (p. 17)

A number of researchers have argued that some of the sustainability discourses are limited, prompting a need to distinguish between different forms of sustainable development and sustainability (e.g., Hopwood et al., 2005). Working beyond music education, researchers have challenged ecological movements as being superficial and focused on feeling good about one's own actions rather than considering how the systems-level problems of sustainability penetrate the whole of life on multiple levels. We ask further, "What is needed from music education to move towards more ecologically aware practices, while operating within an anti-ecological, neoliberal society that is focused on learning outcomes that prioritise the production of economically successful citizens?" One of the obvious tensions for place-based approaches in education is the potential for conflict with the "social imaginaries" of public education. As Barrett (2012) noted:

> Where educational reform in neoliberal societies is "deeply committed to a standards and testing culture that tends to ignore the peculiarities of place in order to standardize the experiences of students" (Graham, 2007, p. 375) ... ecological approaches to education are often ignored, if not actively discouraged. (p. 209)

Consequently, the discourse of ecological sustainability in music education also carries inherent tensions that link to the discourse of the politics of unsustainability. Early approaches to ecological sustainability in our field tended to reduce responses to the complex questions of sustainability to a romanticisation of past musical traditions and practices in the interests of preservation and conservation, without necessarily engaging with the deeply inequitable and unethical practices which attend some of these traditions. In engaging with the deeper educational ethos and forces that steer professional thinking, an ecological focus turns back from "me," "us," and "our field" to ask, "What is music education doing or leaving undone in its ecosystemic environment?" Such a focus shifts us from questions of "How can we preserve music education practice in schools?" to ones of "How can school music education practices serve the community and diminish inequalities?" "What transformation is needed from music education itself?" Without such questions, ecological sustainability in music education risks becoming an ego-logical practice focusing inwardly on the profession's own sustainability and self-justification rather than considering how the profession might create better conditions for justice and fairness in and through music and music education.

This shift towards critically exploring the unintended and undesirable consequences of music education systems and practices turns the sometimes-narcissistic gaze of music education outwards to recognise relationships, horizontal dependencies, and unethical imbalances. We are now prompted to think and create reconfigurations that could address these imbalances. This means cultivating doubt in our belief in limitless growth, a growth which COVID-19 so effectively challenged. In the end we cannot separate professional practice from the wider social effects, from cultural paradigms, and the potentially unrecognised coeffects and synergies of our professional practices on various ecosystems including economies, and human well-being locally and globally. What this means in practice may be something we might never have contemplated but were suddenly forced to during the COVID-19 pandemic.

Education approaches such as EcoJustice Education (Martusewicz et al., 2015) and Freirean Ecopedagogy (Misiaszek, 2017) draw from a critical and ethical analysis of culture to identify "the patterns of belief and behavior in our culture that have led to destructive relationships and practices that have harmed the natural world, as well as human communities" (Martusewicz et al., 2015, p. 10). As do we, EcoJustice Education resists the culture–environment dichotomy while searching for "the cultural roots of the ecological crisis" and advocating "eco-ethical consciousness" (p. 12). Arts educators Mary Ann Hunter and colleagues (2018) have argued that an approach to a more sustainable education lies "not so much in *what* is taught, but *how*" (p. 14). They draw on educational philosopher Maxine Greene's (1995) concept of "wide-awakeness" in their analysis of ecological aspects of arts education as requiring not just "conveying facts about impact and effect" but also recognising "how practices can foster the agency and wide-awakeness" through holistically combining issues of justice, equity, and ecology (p. 14). Hunter and her colleagues prompt further important questions, such as "What is to be sustained? What needs to change and how? Who decides what change or sustainability looks like? What are the implications for practice?" (p. 16). To these questions we add, "What do we need to conserve in music education? What needs to be transformed? What does such transformation require of the profession and its practitioners?".

Towards *Eco*-Logical Thinking in Music and Music Education

What first steps should we take then for approaching a better understanding of how our professional field can move towards ecological thinking and ecopolitical music education? Ecologically oriented arts educators ask questions such as how can we trouble "the grand narratives of economic progress, individualism, and dissociation from our planet's fragile future" (Hunter et al., 2018, p. 1) that dominate our societies? How might we educate teachers to better navigate this complexity, uncertainty, and the moral dilemmas that are involved? Barrett (2012) traced the roots of ecological thinking within music to Gregory Bateson's work, in particular his volume of essays *Steps to an Ecology of Mind* (1972). Like Dewey (1925), Bateson argued that the mind is constituted in and through the systems with which it engages, it is not "in the head" alone. Bateson illustrated this argument through an analysis of features of Balinese "character" and the structures of Balinese music. The anthropologist and music educator Christopher Small drew on Bateson's notions of the ecological mind in his seminal book *Music, Society and Education* (1977) in order to illustrate the interconnectedness of musical thought and action with the social, cultural, and natural worlds in which they occur. Small considered music as "a *process*, by which we explore our inner and outer environments and learn to live in them" (pp. 3–4). He expanded on this view in his proposal of *musicking* first introduced in his 1987 text *Music of the Common Tongue* and fully elaborated in his 1998 text *Musicking*. Rather than taking music merely as something that people do, Small (1998) suggested that it is "an important component of our understanding of ourselves and our relationships with other people and the other creatures with which we share our planet" (p. 13). Small's analysis of musical practices allows us to understand how our social ecologies—including the power relations—are shaped in and through music (also Small, 1987). In musicking, Small (1987) highlighted the active, participatory, and communal nature of musical action and the ways in which it both reflects and shapes society.

This active and generative side of musicking (often missed in interpretations of Small's work) positions the ecosystems of music education at the centre of society and its desired values. By affirming and celebrating who we are in relation to fellow humans and to the world (Small, 1987) music education can engage in issues such as justice and fairness in and through musical practice. In musicking, we can explore our relationships, not simply through transmission of existing ones, but also by rethinking where the imbalances exist. As Odendaal and colleagues concluded, Small's Batesonian ecological view highlights the potential of musicking as a lens to understand what happens in music education as "an educative process in the sense that those who 'music' learn new things of themselves and of the contexts in which they 'music'" (Odendaal et al., 2013, p. 163). This perspective leads directly to issues of justice and fairness, equality and equity. Musicking also prompts us to ask, "Who is this 'we'?," and "What are 'our' most inwardly desired relationships?," "What is the public image of our musicking?" As a caution, while musicking we may celebrate certain relationships that at the same time concretely and continuously exclude some other people.

As did Small, scholars in music education have begun to draw on theorisation of place, in order to shape a more "place-conscious" or "ecologically valid" localised music education. Music educator Sandra Stauffer (2009) argued that music curriculum developers should look to the *sociomusical* practices of the locale in order to develop a "place-conscious music education" that seeks "to reconnect schools and communities and lived experience" (p. 178). In such social–ecological views, music educators might become "community music educators" who view music practices as "fluid, dynamic, and contextual and who recognize the need for continual examination of the intersections of people, place, and practice" (Stauffer, 2009, p. 183). Others have drawn on ecological perspectives in diverse aspects of music, including investigating the relationships, theory, and practices in music, health, and well-being (DeNora, 2011), creative collaborative music practices (Barrett, 2014), adults' free vocal improvisation choir practice (Siljamäki, 2021), and children's ecologies of music practice (Barrett, 2005a, 2005b).

For some music educators, this emerging ecological shift in music education includes a critique of the Western human-centred foci of current conceptions of music education. In his volume *Eco-Literate Music Pedagogy*, Daniel Shevock (2018) suggested a music pedagogy that is place-rooted, connecting music and nature in meaningful and ethical ways, "developing ecological consciousness by ritualizing and creating music rooted in soil" and "connecting to the planet more broadly by connecting local understandings to global ecological crises" (p. 10). For Shevock, the capacity of music and musicking is connected to cultivating *ecological literacy* in order to create consciousness of how deeply embedded the systems of oppression are in relation to issues of gender, race, and class. For some, this is combined with spiritual and specifically Christian aspects of human existence (Bates et al., 2021). Bates et al. (2021) situated their critique in the context of "specific Indigenous North American cultures (e.g., Western Apache, Nuu-chah-nulth, Stó:lō, and Syilx)" to argue that "even the most inclusive and intersectional analyses tend to be *anthropocentric*—centered on human experiences, needs, and desires, regardless of impacts on nonhuman beings and places" (p. 164). They suggested a political ecology of music education in which human diversity is situated:

> within *ecodiversity* (plant and animal life, water, minerals, landforms, weather, and so forth), thereby extending the scope of diversity and justice to include political concerns about climate change, ecosystem destruction, and extinction, while also deepening understandings essential to overcoming human oppression, domination, and exploitation. (p. 164)

Such politics of music education respects localism and "a grassroots 'pluriverse'" (Esteva & Prakash, 2014, in Bates et al., 2021, p. 166) and can "help flatten current hierarchies, human or otherwise, and extend justice beyond the anthropocentric" (p. 165). While a place-based music education seeks to embrace a more holistic nature of music learning and teaching in a particular setting or from the perspective of a particular group of people:

there are inevitable tensions in the ways in which the needs of the local can be addressed and accommodated within the structures and policies of national curriculum policies and practices, and the needs of the larger intersecting cultures of musical practice. (Barrett, 2012, p. 207)

Concluding Remarks

In this chapter we have explored the emerging discourses of ecological thought and action, sustainability policy and practice at global and local levels, and the ways in which these have been taken up recently in music education. A concern for an ecological approach to music education has its roots in the early work of Small and reflects Dewey's long-standing recognition of relationality of the individual, the collective, the environments in which they live and work, and meaning making. We suggest that Dewey's work was prescient in its time—a period that was marked by teleological thinking in response to problems—and informs our approaches to contemporary challenges. An ecopolitical music and music education as outlined in this book implies a rethinking of music education professionalism. We shall explore the implications for a reconceived notion of music and music education professionalism as transformative ecopolitical professionalism in the next chapter.

References

Barnett, R. (2017). *The ecological university: A feasible utopia* (1st ed.). Routledge.
Barrett, M. S. (Ed.). (2014). *Collaborative creative thought and practice in music*. Ashgate.
Barrett, M. S. (2005a). Musical communication and children's communities of musical practice. In D. Miell, R. A. R. MacDonald, & D. J. Hargreaves (Eds.), *Musical communication* (pp. 261–280). Oxford University Press.
Barrett, M. S. (2005b). Children's communities of musical practice: Some socio-cultural implications of a systems view of creativity in music education. In D. J. Elliott (Ed.), *Praxial music education: Reflections and dialogues* (pp. 177–195). Oxford University Press.
Barrett, M. S. (2012). Troubling the creative imaginary: Some possibilities of ecological thinking for music and learning. In D. J. Hargreaves, D. Miell, & R. A. R. MacDonald (Eds.), *Musical imaginations: Multidisciplinary perspectives on creativity, performance, and perception* (pp. 206–219). Oxford University Press.
Bates, V. C., Shevock, D. J., & Prest, A. (2021). Cultural diversity, ecodiversity, and music education. In A. Kallio, H. Westerlund, S. Karlsen, K. Marsh, & E. Saether (Eds.), *The politics of diversity in music education* (pp. 163–173). Springer.
Bateson, G. (1972). *Steps to an ecology of mind*. The University of Chicago Press.
Blühdorn, I. (2017). Post-capitalism, post-growth, post-consumerism? Eco-Political Hopes beyond Sustainability. *Global Discourse, 7*(1), 42–61.
Blühdorn, I., & Welsh, I. (2007). Eco-politics beyond the paradigm of sustainability: A conceptual framework and research agenda. *Environmental Politics, 16*(2), 185–205. https://doi.org/10.1080/09644010701211650
Bronfenbrenner, U. (1974). Developmental research, public policy, and the ecology of childhood. *Child Development, 45*(1), 1–5.

References

Bronfenbrenner, U. (1977). Toward an experimental ecology of human development. *American Psychologist, 32*(7), 513–531.

Darling, F. F. (1951). The ecological approach to the social sciences. *American Scientist, 39*(2), 244–254.

DeNora, T. (2011). *Music-in-action. Selected essays in sonic ecology*. Routledge.

Dewey, J. (1938/1986). Logic: The theory of inquiry. In J. A. Boydston (Ed.), *John Dewey: The later works 1925–1953* (Vol. 12, pp. 1–528). Southern Illinois University Press.

Dewey, J. (1925). *Experience and nature*. Open Court.

Dyball, R., & Newell, B. (2015). *Understanding human ecology*. Routledge.

Feeney, S. (2020). Transitioning from the MDGs to the SDGs: Lessons learnt? In S. Awaworyi Churchill (Ed.), *Moving from the millennium to the sustainable development goals* (pp. 343–351). https://doi.org/10.1007/978-981-15-1556-9_15.

Graham, M. A. (2007). Art, ecology and art education: Locating art education in a critical place-based pedagogy. *Studies in Art Education, 48*(4), 375–391.

Greene, M. (1995). *Releasing the imagination: Essays on education, the arts, and social change*. Jossey-Bass.

Hopwood, B., Mellor, M., & O'Brien, G. (2005). Sustainable development: Mapping different approaches. *Sustainable Development, 13*, 38–52.

Hunter, M. A., Aprill, A., Hill, A., & Emery, S. (2018). *Education, arts and sustainability. Emerging practice for a changing world*. Springer Briefs for Education.

Keller, D. R., & Golley, F. B. (Eds.). (2000). *The philosophy of ecology: From science to synthesis*. University of Georgia Press.

Maris, V. (2015). Back to the Holocene. A conceptual, and possibly practical, return to a nature not intended for humans. In C. Hamilton, F. Gemenne, & C. Bonneuil (Eds.), *The Anthropocene and the global environmental crisis: Rethinking modernity in a new epoch* (pp. 123–133). Taylor and Francis.

Martusewicz, R. A., Edmundson, J., & Lupinacci, J. (2015). *EcoJustice Education: Toward diverse, democratic, and sustainable communities* (2nd ed.). Routledge.

Misiaszek, G. W. (2017). Educating the global environmental citizen. *Routledge*. https://doi.org/10.4324/9781315204345

Morton, T. (2018). *Being ecological*. The MIT Press.

Odendaal, A., Kankkunen, O.-T., Nikkanen, H., & Väkevä, L. (2013). What's with the K? Exploring the implications of Christopher Small's "musicking" for general music education. *Music Education Research, 16*(2), 162–175. https://doi.org/10.1080/14613808.2013.859661

Plumwood, V. (2001). *Environmental culture: The ecological crisis of reason*. Routledge.

Shevock, D. J. (2018). *Eco-literate music pedagogy* (1st ed.). Routledge.

Siljamäki, E. (2021). *Plural possibilities of improvisation in music education: An ecological perspective on choral improvisation and wellbeing*. [Doctoral dissertation]. Studia musica 86. Sibelius Academy. University of the Arts Helsinki.

Small, C. (1977). *Music, society and education*. John Calder.

Small, C. (1987). *Music of the common tongue: Survival and celebration in African American music*. Wesleyan University Press.

Small, C. (1998). *Musicking: The meanings of performing and listening*. Wesleyan University Press.

Stauffer, S. (2009). Placing curriculum in music. In T. Regelski & J. T. Gates (Eds.), *Music education for changing times: Guiding visions for practice* (pp. 175–186). Springer.

United Nations Development Program (UNDP). (2022). *The SDGs in action*. https://www.undp.org/sustainable-development-goals?c_src=CENTRAL&c_src2=GSR. Accessed 12 July 2022.

Washington, H. (2015). *Demystifying sustainability: Towards real solutions*. Routledge.

Chapter 3
Rethinking Professionalism in Music Education

Abstract Professionalism is a contested concept as it responds to various socio-political agendas, shifting perspectives, competing ideologies, and power issues not revealed in common understandings of the term. This chapter interrogates the assumptions that underpin professionalism in music education in order to propose an ecopolitical professionalism that drives systems transformation when addressing "wicked problems." The chapter critiques the "technical rationality" that has shaped much music education theory and practice through an uncritical emphasis on maintaining musical traditions and established methods, and presents a view of music educators as transformative systems thinkers and practitioners in the moral ecologies of the field. Ecopolitical professionalism requires the use of "moral imagination" to generate social and societal transformation through public pedagogy.

Developing Ecopolitical Music Education Professionalism Within—Beyond Technical Rationality

As noted in the introduction to this volume, professionalism is a contested concept as it responds to various socio-political agendas, shifting perspectives, competing ideologies, and power issues not revealed in common understandings of the term (Cribb & Gewirtz, 2015, p. 23). As Andrew Abbott noted in his seminal study *The System of the Professions* (1988), such contestation is not purely a contemporary issue; rather, it has been a feature of professions historically. In this chapter we consider the assumptions that underpin professionalism in music education in order to propose ecopolitical professionalism as "a powerful instrument of occupational change" (Evetts, 2014, pp. 34–35) that might be drawn on as musicians and music educators work responsively in changing contexts and conditions. We seek to illuminate the potentialities and emergent manifestations of an ecopolitical professionalism that engages with wicked problems. As with professionalism, ecopolitics may be viewed as a contested term as its concerns are interrelated and expanding rather than clearly bounded. To avoid relational excesses, we therefore apply the principle of the "ecology of separation," following Esposito's (2017) proposition to radicalise and

© The Author(s), under exclusive license to Springer Nature Switzerland AG 2023
M. S. Barrett and H. M. Westerlund, *Music Education, Ecopolitical Professionalism, and Public Pedagogy*, SpringerBriefs in Education,
https://doi.org/10.1007/978-3-031-45893-4_3

sharpen (rather than blur) relevant distinctions. Such work invites music educators to consider themselves as transformative systems practitioners whose work responds to the embedded challenges of education and socio-cultural systems. We conclude this chapter by considering what it means to think and act within ecological systems and engage in moral and ethical acts of public pedagogy.

Biesta (2017) identified three assumptions of professionalism, those of: a concern for "human well-being," requiring "highly specialist knowledge and skills," and work in "relationships of authority and trust" (p. 318). He suggested that these assumptions can provide both a definition of professionalism as well as a justification, and points to the risk that such assumptions lead to arguments for levels of qualification and accreditation that "ensure" quality and universal standards and create closed, inward looking, and self-regulatory entities. In this critique of professionalism, it can thus be seen as a process of "closing" the market, a process of claiming space and "defining boundaries of service and expertise and then gaining control of the territory thus created" (Fenwick, 2016, p. 23). Music therapist and music educator Taru-Anneli Koivisto (2022) provided a compelling example of this through her account of the ways in which music therapy has demarcated its territory in many countries through developing professional education and accreditation standards that define the boundaries of service in relation to other professions including music education.

While navigating between different conceptions of professionalism, we acknowledge that there is a risk that too rigid a definition asserts one perspective over others. Yet, some conceptual closures may be necessary in order to create an understanding of the phenomenon that we would like to highlight. We agree with researchers of professionalism Alan Cribb and Sharon Gewirtz (2015) in their view that, in general, "professionalism" is considered to be a kind of "team sport" that is more collective than self-oriented. In this relational sense, professionalism implies the use of one's expertise (whether or not this is formally recognised and endorsed) in the interests of wider society.

Several conceptual boundaries can be identified in understanding ecopolitical professionalism in music education. In the following, we explore the affordances and constraints these offer.

First, we see ecopolitical professionalism in music education as a form of professionalism that can be developed "from *within*" (Evetts, 2014, pp. 40–41) thereby avoiding the often strong resonances of standardisation and managerialism related to professionalism from above. Our view can be seen as an attempt to rebalance the power between professionals and organisational management (Cribb & Gewirtz, 2015, p. 33). We also emphasise responsibility instead of occupational status and monetary benefits, the latter often associated with professionalism. Professional autonomy (Hargreaves, 2000) becomes central to this rebalancing as music teachers exercise the capacity to develop their work with a reflexive and transformative mindset while at the same time avoiding potentially detrimental individualism and adherence to a silo mentality. Thus, professionalism refers to agentic action, "not to feel obliged to jump through hoops devised by others but to be able to steer one's own course by making one's own independent judgements" (Cribb & Gewirtz, 2015, p. 50).

Second, we highlight the "artistry" in ecopolitical professional work when combined with ecological concerns. In this sense ecopolitical professionalism expands its interest areas beyond the usual to embrace imaginative possibilities of thought and action. In everyday discussion, music education professionalism tends to be restricted to considerations of musical and pedagogical expertise and its related professional education. This "technical rationality" (Schön, 1983/2016, pp. 21–37) emerged in understanding professionalism after World War II as a response to the need to rebuild societies (Solbrekke & Sugrue, 2011; Sullivan, 2005) and increased hierarchical differentiation of professions (also identifiable in the music field in specialised professions such as music technology expert, music psychologist, music therapist). In occupations in general this development created particularism and the consequential "limited responsibility defined by local circumstances and the problem to be solved" (Solbrekke & Sugrue, 2011, p. 16) as a socially and politically neutral technical rationality was developed "at the expense of moral and societal responsibilities" (p. 16). A historical examination has hence identified a shift among professionals from so-called "social trustee professionals" to "expert professionals" (p. 16). This shift in self-understanding has contributed to "the rise of a utilitarian ethos" (Brint, 2002, p. 245), manifest in "a tendency to think of professional education merely as a means of obtaining credentials that will be valuable for each individual on the labour market" (Solbrekke & Sugrue, 2011, p. 16).

In educational discourse, technical rationality has been based on rule-following that, according to educational theorist David Carr (2014), fails to cultivate teachers' capacity to identify the morally principled systemic connections between educational activities and wider societal concerns. Schön (1983/2016) explained that in professional practice we face confused ends and conflicting paradigms, and in such a situation there is no rule to be applied, "no clearly established context for the use of technique," and "as yet no 'problem' to solve" (p. 41). That kind of situation requires that both the ends and the possible means are organised "through the non-technical process of framing the problematic situation" (p. 41):

> When we set the problem, we select what we will treat as the "things" of the situation, we set the boundaries, we set the boundaries of our attention to it, and we impose upon it a coherence which allows us to say what is wrong and in what directions the situation needs to be changed. Problem setting is a process in which, interactively, we *name* the things to which we will attend and *frame* the context in which we will attend to them. (p. 40)

In music education, technical rationality derives its main artistic, pedagogical, and intellectual principles and values from musical traditions and established methods. Technical rationality in music therefore tends to be oriented towards the past, referencing world-orders that once seemed justified. It is reflected in a focus on musical works or genres, thereby de-politicising the professional field by narrowing its complex situational and contextual ethics to a loyalty to the principles of a (narrow) range of musical heritage and established practices. Technical rationality sustains a hierarchy of values in which past artistic values dominate, with anything outside these considered to be of lesser value. It often also relegates social values to a lower status and, in that process, ignores the fact that music education is always also a social system.

In organisational contexts, technical rationality does not acknowledge the intersubjective and collaborative potential of professional work and professional education. Rather, it prevents professionals from recognising and creating other lenses to identify the relationship between their own expertise and the social realities of society. In this way, as technical rationality creates a false assumption of societal and political neutrality it prevents professionals from seeing how their own organisations sit within larger systems in society (Senge et al., 2008). Further, false assumptions of societal and political neutrality can lead to a form of professional imperialism (Midgely, 1981). James Midgely introduced the concept of professional imperialism to the field of social work in order to highlight the ways in which settler, colonist theories and practices were introduced uncritically to countries in the majority world without consideration of local and Indigenous knowledges, values, and practices. Within music education, the uncritical introduction of various methods of music instruction might also be seen as a form of professional imperialism.

Third, when technical rationality underpins professional work, music educators are not encouraged to draw on and use their own social imagination (Greene, 1995) to explore the potential for novel relational musical spaces to generate new relationships and open dialogues (Barnett, 2015, 2018). Technical rationality approaches focus on producing the right answers instead of looking at things "as if they could be otherwise" (Greene, 1995, p. 19). More importantly, since the technical rationality of professions distances itself from moral and societal responsibilities and experimentation, effective action consists in maintaining "business as usual." Technical rationality tends to valorise testing, grading, individualistic curricular goals, and atomistic standards, as well as competitive learning environments that operate with musical criteria for quality, in this way further bolstering an ego-logical understanding of, and research on, what music is *for* in education. This kind of technical rationality is deeply rooted in the modern school system and the music teacher education that serves this school system, and frames and legitimises professional practice and education in music beyond the school context (e.g., extracurricular instrumental teaching). Within this rationality it is justified to say that wider societal questions are not of my or our concern. "I am just a musician!" Or, "I am just teaching music!".

Fourth, instead of seeing ecopolitical professionalism as a field of technical certainties, we see the judgements of music educators as involving values and being informed by educational ideals (Biesta, 2009). We identify music education professionalism as a *process* replete with uncertainty and complexity, filled with tensions between ideological control and normative value, yet leaving space for individual music educators to work against prevailing ideological controls in their respective environments (Fenwick, 2016). We therefore agree with Schön (1983/2016) that, instead of trying to make professional work simpler, there is a need to see it as a "swamp." Schön (1983/2016) wrote:

> In the varied topography of professional practice, there is a high, hard ground where practitioners can make effective use of research-based theory and technique, and there is a swampy lowland where situations are confusing "messes" incapable of technical solution. The difficulty is that the problems of the high ground, however great their technical interest, are often relatively unimportant to clients or to the larger society, while in the swamp are the problems

of greatest human concern. Shall the practitioner stay on the high, hard ground where he can practice rigorously, as he understands rigor, but where he is constrained to deal with problems of relatively little social importance? Or shall he descend to the swamp where he can engage the most important and challenging problems if he is willing to forsake technical rigor? (p. 42)

In our current complex world, professionalism unavoidably involves tensions and struggles that demand a readiness to see professional work "in the context of broader debates about social and civic purposes" (Cribb & Gewirtz, 2015, p. 71) so that "tricky ethical and political dilemmas should properly be seen as falling *within* rather than *outside* the remit of professional ethics" (p. 73; emphases added). Rather than holding ever more tightly to the "high ground of technical rationality," we need to recognise that wicked problems cannot be tamed (Ison, 2017, p. 36), but need to be engaged with continuously, in the "swamp": As Ray Ison stated, "Our circumstances have become such that more of the same, a business as usual approach, even if done better is no longer good enough" (p. 12).

Fifth, while aiming to reveal the interconnectedness of concerns, ecopolitical professionalism resonates with many contemporary advancements in teacher professionalism, for instance democratic professionalism, collaborative professionalism, and activist professionalism. Dzur (2008) proposed a model of democratic professionalism that is "inherently collaborative" (p. 130). He described it as:

> sharing previously professionalized tasks and encouraging lay participation in ways that enhance and enable broader public engagement and deliberation about major social issues inside and outside professional domains. The sharp distinctions between professional and layperson found in other models disappears here. This does not mean that professional authority, status, privilege, and responsibility disappear as well, only that they are tightly connected to the empowerment of laypeople. (p. 130)

This description emphasises mutuality, recognising the complementary knowledge that all parties bring, and connecting all in a collaborative approach to problem solving. Collaborative professionalism (Hargreaves & O'Connor, 2015), for its part, emphasises not just professional collaboration but also shared values, concerns, responsibilities, and complementary skills sets and understandings harnessed to co-create better practice. Activist professionalism (Sachs, 2003) is yet another name for transformational professionalism. Writing in 2003, Sachs described an activist teaching profession as "an educated and politically astute one" (p. 154). Initially framed as a "moral urging," Sachs (2016) reflected in more recent writing on the "general acceptance that a different kind of professionalism is needed," one that moves beyond an activist professionalism that is inwardly focused to one that "requires that teachers collectively and individually address those in power to make it clear that a top-down approach is simply not working" (p. 414). These articulations for transformative professionalism each work as important conceptual counterforces not only to past technical rationality in expert work but also to managerial professionalism that diminishes music teachers' autonomy as change agents in society.

We suggest that ecopolitical professionalism in music and music education might serve as an overarching conceptual frame for democratic, collaborative, and activist professionalism, which works both inwards and outwards, serving not only the individual concerns of the teacher but also providing a framework for our professional practice to address the wicked problems we now face. The professionalism that we advocate therefore involves "being routinely reflexive and self-critical" and ready to play an active role in redefining the managerial and organisational norms, as well as "being prepared to work both inside and outside employing organizations to help shape the broader social landscape" (Cribb & Gewirtz, 2015, p. 72). However, as Cribb and Gewirtz (2015) explain, "this is not to suggest that professionals should always see themselves in combative or antagonistic relationships with their employing institutions or engaged in overt politics or activism—only that these possibilities need to be embraced within *a full conception of professionalism* [emphasis added]" (p. 72). Such a full conception of professionalism operates transformatively at multiple levels, including those of the self, the discipline, the system, the local, and the global.

Music Educators as Transformative Systems Practitioners

We see the potential of musicians and music educators being and becoming transformative systems practitioners who draw on individual and collective moral imagination to effect ecopolitical change and engage in acts of public pedagogy. Music education is then seen not simply through the musical practices with which teachers and practitioners engage, but also through the field's operational environments as social systems. Above we suggested that the ecological turn to local places, inclusion of Indigenous musics, and practices of music-making is not yet enough (see Chap. 1). Instead, we suggest that a wider social-ecological *systems thinking* (e.g., Biggs et al., 2022; Gonzales, 2020; Ison, 2017) is necessary in order for the field to be able to reveal the workings of music education, its necessary changes, as well as its transformative potential in society. A systems view can thus provide an effective analytical view to examine what music education does and does not do in its environment.

What then is a music education system? Education itself can be seen as a complex and multidimensional system within a larger system, the society, "made up of individuals connected via different social units such as families, schools, communities, religious institutions, business corporations and industries, which interact to fulfill different purposes" (Gonzales, 2020, p. 33). Social systems are contextual, since "every social system is linked to its cultures, traditions and religions, which make up the cultural system within the social system" (p. 33). Yet, as mentioned in Chap. 1, social systems can be identified by their boundaries; those that determine how the system functions and define the components that can be managed by the "system's owners." In other words, music education as a social system can be identified by what the practitioners see as its purpose and what its purpose is not. A systems boundary refers to "the periphery or bounds within which systems' components work together—in other words, to the limits that identify the system's components, processes and interrelationships when they interface with another system" (p. 5).

It is relatively rare in music education to look at practices from the perspective of social-ecological systems; rather, the discipline typically positions the individual learner in relation to a more or less abstract musical practice in which musical agency is gained. Such learning/doing is rarely examined through the multiple lenses of the surrounding contexts, including the social and institutional contexts, consideration of who is included or excluded, and/or what the practice "does" and does not do in its wider ecosystem. The typical narratives for the "purpose" of music education tend to be individualistic, limiting the horizons for professional responsibility within the boundaries of the established and known practice and in that process, reinforcing the technical rationality of professionalism. Yet, we know that professional work in music education is not immune to challenges that can be seen as systems challenges rather than, for instance, problems of teaching per se. Complex social problems such as injustice, exclusion, or inequalities can become embedded in music education social systems when they are selective rather than all-inclusive systems (Ilmola et al., 2021; Väkevä et al., 2017, 2022; Westerlund et al., 2021).

Instead of analysing how to engage these questions, there is a tendency for music education to be silent on systems problems and align with the discourse of the benefits of music and positive impact of music education. Ethnomusicologist Geoff Baker (2021) analysed this "ambivalence about music and the ambiguity of its effects" by arguing that the narratives of the positive potential of music "only [tell] part of the story" (p. 9). These positive impact narratives may be seen as necessary to secure funding and boost self-esteem, but narratives that "exaggerate benefits, elide ambiguities, and minimalize problems increase the probability of an education shaped by illusory beliefs rather than rigorous thinking on socially oriented music education" (p. 18). Indeed, music educators have for a long time paid attention to the lack of systems transformation in music education (Colwell, 2012). Instead of limiting the professional perspectives to assumed, hoped for, or aspirational good impacts, it has become necessary to examine these other parts of the story. We are challenged to step up a level to see how each and every music education context operates as a social system, what this system's environment is like, and understand the professional responsibility that accrues from this analysis. We then can also imagine how the system needs to be transformed in order for it to be able to transform society.

In professional life, a wider systems thinking approach is increasingly recognised as a key component of both organisational and professional change (Alexander et al., 2019; Gonzales, 2020; Hunter et al., 2018; Westerlund et al., 2021). A systems approach "requires a shift in mindset and the ability to consider issues and problems at the micro, meso, and macro levels of the whole system" (Gonzales, 2020, p. 52). It requires moving from a simple additive approach to curriculum innovation to one which interrogates all levels of thought and action in order to re-structure, re-orient, and make evident the hidden systems of thought and practice as well as uncover hitherto unused possibilities. Systems thinking contributes to music and music education professionalism through fostering:

- an ability to see and understand things from different angles and recognise a range of perspectives that are different from our own (Gonzales, 2020, p. 3), to move between multiple levels, the local and global, near and far, in order "to understand the interplay between the various aspects of the situations … and how the parts of a social-ecological system drive change in each other" (Dyball & Newell, 2015, p. 12);
- "commitment to imagining and taking action for life beyond currently living generations and addressing how humans in the present best sustain future life in a multispecies, more-than-human world" (Hunter et al., 2018, p. 76);
- making visible the work of social processes in order to illuminate the unintended consequences of cycles and connections between seemingly separate actions (such unintended and unwanted consequences are for instance elitism, exclusion, or ego-centric ethos when looked from the perspective of the "big picture" (Ilmola-Sheppard et al., 2021) and pushing "the unawareness of unintended consequences" (Morton, 2018, p. 16) to the forefront of our thinking in order for collective decision making and problem solving to emerge.

In principle, all music educators can be and become *systems practitioners*. A systems practitioner "is anyone managing in situations of complexity and uncertainty—it is not a specialist role or that of a consultant or hired 'intervener'" (Ison, 2017, p. vi). However, literature increasingly identifies how difficult it is for a practitioner to cross boundaries; it is often easier to limit one's professional views to the bounded traditions or taken-for-granted institutionalised practices (Akkerman & Bakker, 2011).

Educational researchers Trevor Gale and Tejebe Molla (2016) have explored notions of "deliberative professionalism" as reflexive action, arguing that "Professional work involves a degree of 'artistry' (Schön, 1983/2016) in bringing knowledge and skills from a particular epistemic culture together with insight into the particularities of a given situation, to inform practice" (p. 251). They identified four models of professionalism: the effective, reflective, enquiring, and transformative professional. The *effective professional* has "little to no recognition of contextual differences or choice in what practices to employ, while the expertise of knowledge domains and professional communities is overlaid, even usurped, by an explicitly political dimension" (p. 251). The work of effective professionals "corresponds with government standardisation of professional practice aimed at achieving national priorities" (p. 251). For instance, music teachers may limit their professional reflexivity to how well they follow the national curriculum guidelines. In teacher education, such effective professionals may resist exceptions—the swamp—and emerging new practices that seem to shake the foundations of their known educational system. The *reflective professional* has more capacity than the effective professional for "carefully considering the particulars of a context in order to discern what techniques should be applied" (p. 251). Gale and Molla (2016) described this approach as a clinical–practice model that draws on the medical concept of triage: "an initial assessment and prioritisation of 'problems' that are then addressed by following standard procedures" (p. 252). The *enquiring professional* is "not just a user of expert knowledge and skills but also a producer of them ... [as] part of an enquiring professional community with which they share the results of their research deliberations" (p. 252). Finally, by combining features from the two previous models, the *transformative professional* is "committed to enquiry that contributes to change, not just new understanding" (p. 252), thus involving a moral and "activist" dimension in their work. Such reflexivity goes beyond reflecting existing practice through the established criteria by shaking up the taken-for-granted understandings pertaining to music education professionalism and professional education.

We argue that this kind of reflexivity requires not just abstract knowledge but also investment in craft. Although ecopolitical professionalism can be seen as a constantly changing and highly contingent frame for transformative professional work in music and music education, a transformative profession can also be seen as a "cultural form" (Spillman & Brophy, 2018) in which craftsmanship and "the desire to do a job well for its own sake" is cultivated (Sennett, 2008, p. 9). The challenge of ecopolitical professionalism lies therefore in how professionals can combine systems thinking, use of abstract knowledge, and enduring craft.

But why is such transformative change still rare? Gale and Molla (2016) emphasised that to be able to work as a transformative professional requires investing time, being exposed to opportunities in which to deliberate and be deliberative, and being challenged to critically reflect on the inequities of social, political, and economic arrangements (p. 259). According to Mabel Gonzales (2020), to engage effectively in systems thinking we must first understand our own thought patterns, values, opinions, biases, and "how we know what we know (metacognition)" (p. 26). Ison (2017), for his part, explained that "the place to start is with the situation and not the system" and that "'distinguishing, or bringing forth' a system in a situation is a particular way of knowing the situation (and that no systems exist a priori)" (p. 49). Systems thinking may thus produce heuristic devices that can be helpful in exploring a situation and to deliberate with alternatives rather than offer fixed solutions.

We acknowledge that a transformative ecopolitical professionalism in music and music education requires crossing boundaries within and beyond the system thereby facilitating transactions beyond some gap or barrier. If systems transformation requires understanding of how the organisation is functioning, what its purpose is, and what is and what is not within the boundary of the social system, a systems boundary in itself is a conceptual attempt to understand the system in its environment. For instance, if the purpose of music education is to maintain a professional music ecology through selecting the most talented children for specialist music instruction, then it may be impossible to see how exclusion takes place, or to recognise how the music education system perpetuates inequities and exclusion and sustains opportunity gaps. The rhetoric of "music for all" in such settings precludes consideration of who the "all" encompasses and ignores the agency and voice of children and their interests (Barrett, 2017) and the informal and non-formal places and spaces in which music learning and engagement occur (Barrett & Westerlund, 2017).

Becoming aware of and working towards establishing linkages with organisations beyond the systems boundary has been described as "boundary spanning" (Gonzales,

2020, p. 55; see also Akkerman & Bakker, 2011). Transformative systems practitioners as boundary spanners are individuals who facilitate transactions between people or groups beyond organisational silos, professional "tribes," and between ecosystems (Folke, 2006). Boundary spanners cross organisational boundaries, such as departments or organisations or disciplinary cultural boundaries, in order "to exchange knowledge or mediate interactions" (Long et al., 2013, p. 2). Boundary spanning opens new possibilities to contest prevailing discourses and transgress hidden cultural structures. We argue that such work demands a form of moral and social imagination; that is, a capacity to consider the moral ecologies of the intersecting systems in which music educators work, to embrace the transformative possibilities of ambiguity and complexity and engage in public pedagogy.

Expanding Moral Ecologies of Music Education Through Public Pedagogy

If we view a social system as a complex network of human relationships (individuals, groups, institutions), change of and to a social system can only take place through people; those who have the power to generate new social worlds. Such work requires of individuals and groups a capacity to exercise their agency and reposition their actions, beliefs, and values in response to emergent and new demands of the context. The ecopolitical professionalism we wish to advance takes a processual approach to transformation; one that acknowledges that change is continual and non-linear and embraces pluralism and the potential for conflicts between values. The processual view moves beyond a dualistic approach in which there are strictly speaking only intrinsic and instrumental values in music and music education (see also relational values literature, including Himes & Muraca, 2018; Saxena et al., 2018; Abbott, 2016). The processual view is necessary in order to understand the complexity of experience and the intertwined nature of meanings as well as the ways relationships between our doings matter not just as a means to certain ends, but as ends in themselves. In other words, "qualitatively good ends are not absolutes that lie beyond activities but the ends are constituted by the activities themselves" (Westerlund, 2008, p. 83). The American social theorist, Andrew Abbott (2016), drew on Dewey's idea of the means–ends continuum when he wrote:

> In the processual view, the means/ends distinction falls apart. It can be made at any given time, but it is meaningless across time. Every end is both a moment of consummation and a starting point for new projects. Every present becomes past, every end disappears. No end can be permanently secured, because even the past is being perpetually rewritten in line with the present, and will be again rewritten in the future. In any case, there is no unchanging self with respect to which means/end status can be decided. The whole means/ends distinction comes apart when viewed diachronically. (p. 282)

Values as acted and experienced cannot therefore be judged by simply evaluating the musical outcome with unchanging criteria. The social life is made and remade pointing towards the "potential for action and the openness of the present to genuine choice" (Abbott, 2016, p. 230). In this sense, all social life—and music education as part of social life—is moral: "the moral issue concerns the future: it is prospective …. The moral problem is that of modifying the factors which now influence future results" (Dewey, 1922, p. 18, as cited in Abbott, 2016, p. 272).

Sociologist Zygmunt Bauman (1993) argued some decades ago that contemporary morality must differ from past eras of ethical understanding and practice in life and professional work. Bauman (2011) noted that in former eras of "codified morality" the application of ethical norms, codes, and laws was assumed to guarantee the "right" results. He asserts that we have now entered "an epoch of agreement to the permanent coexistence of diversity" (p. 93). This means the quest for a renaissance of morality and responsibility requires that we are able to "rethink the trends of the past" (Bauman & Donskis, 2013, p. 211). For Bauman, responsibility calls for a personal stance that is conscious of its obligations to others, even in the face of neglect, opposition, or ignorance. Responsibility means that the first reality of the self is to be *for* the Other before one can be *with* the Other (Bauman, 1993):

> To take a moral stance means to assume responsibility for the Other; to act on the assumption that the well-being of the Other is a precious thing calling for my effort to preserve and enhance it, that whatever I do or do not do affects it, that if I have not done it it might not be done at all, and that even if others do or can do it this does not cancel my responsibility for doing it myself. … being-for is unconditional. (Bauman, 1995, pp. 267–268).

This kind of relational morality that Bauman seeks becomes more complex when there is a need to take responsibility for our "responsibility for otherness" for many (Bauman, 1993, p. 130). When we accept and understand that moral dilemmas cannot be solved by laws and codes, we arrive at a position where there cannot be a codified ethics. Thus, there is no one right predefined choice to be followed in each and every situation and context. In this kind of relational morality there are no universal recipes. Rather, we must wake up our moral conscience, start nurturing our sense of professional responsibility, and bear the discomfort and necessity of constant struggle "when reflecting what responsibility means" (Westerlund, 2019, p. 513).

Moral ecology as a concept aims to capture the way multiple and even contradictory normative discourses are negotiated in working environments and highlights how organisational and professional environments are "moral forces that shape judgment and behavior in their own right" (Plaisance, 2021, p. 434).[1] The moral ecology of environments (such as classrooms) "have a bearing on the manifestation of virtuous behavior" drawing "attention to the structures and norms that mediate among various needs and interests" (p. 427). Identifying the moral ecologies of environments prompts questions such as:

- How might the moral ecologies of music education be transformed in order to respond to major social changes and the consequential exclusions, inequalities, and injustices?

---

[1] The term moral ecology has been used interchangeably with human ecology, social ecology, or community ecology, yet focusing on moral ecology sharpens the theoretical "focus on the general rather than the relativistic" (Hertzke, 1998, p. 637).

- What kinds of moral reflexivity and reimagination are then needed from music educators?
- How can relational morality actualise a music education that aims towards ecological awareness?

Music education that is underpinned by technical rationality attempts to answer such questions having arrived at firm rules and principles, and, in this way, aims to reduce uncertainty, negotiation, and ambivalence. It is increasingly evident that we need to search for more future-oriented, complex, contextual, and situational ethical perspectives. Transformative ecopolitical professionalism in music and music education can be seen as an aspirational and imaginative process, "a relation between the habits of individual professionals and professions and the changing world around them" (Abbott, 2016, p. 274). Placing morality at the centre of ecopolitical professionalism requires therefore "moral imagination" (Bauman & Donskis, 2013) that can be seen as "the free play of thought when entertaining choices" and the "capacity to see a matter as if it were otherwise" (Allsup & Westerlund, 2012, p. 140). As eminent educational philosopher Maxine Greene (1995) wrote:

> To tap into imagination is to become able to break with what is supposedly fixed and finished, objectively and independently real. It is to see beyond what the imaginer has called normal or "common-sensible" and to carve out new orders in experience. (p. 19)

In this book we suggest that public pedagogy provides a means to expand the moral ecologies of music and music education. We consider the potential of public pedagogy in music education as work that makes space for interrelations and resonances between politics, arts, and education, and reopens the public sphere for participants "beyond pedagogy as a cognitive and rational process of transmission" (Sandlin et al., 2017, p. 827). We recognise a need for the field to seek to move beyond self-advocacy towards ecopolitical awareness and action: from recognition of issues and public admissions of complicity in supporting and maintaining continuing inequities to the development and implementation of practices that restructure the systems to which we have become accustomed. In short, to ask better questions and arrive at better solutions.

The call to public pedagogy is not new. Allsup and Shieh (2012) advocated for music educators in school settings to reach "into larger and more intertwined social, artistic and political domains" (p. 47) in order to engage in public pedagogy in relation to social justice. Others have examined protest music as a form of public pedagogy in musicological analyses (Haycock, 2015), while the role of music in popular culture practices beyond the school and university institutions has been long recognised as a component of public pedagogy (Sandlin, 2009). In this discussion we point out the differing understandings we can have for and of musical performance. Whereas music education typically sees performance from the perspective of cognitive musical knowing-in-action, we suggest that music performance as public pedagogy in music and music education can manifest social transformation and musical performance as political place-making. For a music educator public pedagogy can thus become an arena for rethinking and reimagining the moral ecologies of music and music education. In considering public pedagogy as a dimension of transformative ecopolitical professionalism we thus seek new conceptions of ecology, ecological thinking, and emerging commitments to ecological practice in music education in order to understand our professional responsibilities in the public sphere to the moral ecologies in which we live and work. This approach recognises that we live and work in processual worlds, constantly being made and remade, where we become transformative systems practitioners in public.

Concluding Remarks

In this chapter we have made a call for rethinking professionalism in music education and to critique the technical rationality that has underpinned much music education theory and practice. We have argued that this might be achieved through developing a transformative ecopolitical professionalism which positions music educators as transformative systems practitioners who also recognise the transformative potential of public pedagogy in the field and the political potential of music performance beyond learning and knowing. We suggest that through developing an ecopolitical music professionalism we expand the moral ecologies of our work through engaging in public pedagogy. In the following three chapters we present narrative accounts of the public pedagogy work of three individuals, Tuulikki Laes, Riju Talahadur, and Ricky Kej. These narrative accounts illustrate the processes and outcomes of engaging in transformative ecopolitical professionalism at local through to global levels, across intersecting social-ecological systems, and cultural and political contexts around the globe.

References

Abbott, A. (1988). *The system of professions: An essay on the division of expert labor*. University of Chicago Press.
Abbott, A. (2016). *Processual sociology*. University of Chicago Press.
Akkerman, S. F., & Bakker, A. (2011). Boundary crossing and boundary objects. *Review of Educational Research, 81*, 132–169.
Alexander, C., Fox, J., & Gutierrez, A. (2019). Conceptualising teacher professionalism. In A. Gutierrez, J. Fox, & C. Alexander (Eds.), *Professionalism and teacher education: Voices from policy and practice* (pp. 1–23). Springer Nature.
Allsup, R., & Shieh, E. (2012). Social justice and music education: The call for a public pedagogy. *Music Educators Journal, 98*(47), 47–51.
Allsup, R., & Westerlund, H. (2012). Methods and situational ethics in music education. *Action, Criticism, and Theory for Music Education, 11*(1), 124–148.
Baker, G. (2021). *Rethinking social action through music: The search for coexistence and citizenship in Medellín's music schools*. Open Book Publishers.
Barnett, R. (2015). The time of reason and the ecological university. In P. Gibbs, O. H. Ylijoki, C. Guzman-Valenzuela, & R. Barnett (Eds.), *Universities in the flux of time* (pp. 121–134). Routledge.
Barnett, R. (2018). *The ecological university: A feasible utopia*. Routledge.

Barrett, M. S. (2017). Policy and the lives of school-aged children. In P. Schmidt & R. Colwell (Eds.), *Policy and the political life of music education* (pp. 175–190). Oxford University Press.
Barrett, M. S., & Westerlund, H. (2017). Music education in the global context. In G. Barton, & M. Baguley (Eds.), *The Palgrave handbook of global arts education* (pp. 75–89). Palgrave MacMillan. https://doi.org/10.1057/978-1-137-55585-4_575.
Bauman, Z. (1993). *Postmodern ethics*. Wiley-Blackwell.
Bauman, Z. (1995). *Life in fragments: Essays in postmodern morality*. Wiley-Blackwell.
Bauman, Z. (2011). *Culture in a liquid modern world*. Polity.
Bauman, Z., & Donskis, L. (2013). *Moral blindness: The loss of sensitivity in liquid modernity*. Polity.
Biesta, G. (2009). Sporadic democracy: Education, democracy, and the question of inclusion. In M. Katz, S. Verducci, & G. Biesta (Eds.), *Education, democracy, and the moral life* (pp. 101–112). Springer.
Biesta, G. (2017). Education, measurement, and the professions: Reclaiming a space for democratic professionality in education. *Educational Philosophy and Theory, 49*(4), 315–330. https://doi.org/10.1080/00131857.2015.1048665
Biggs, R., de Vos, A., Preiser, R., Clements, H., Maciejewski, K., & Schlüter, H. (Eds.), (2022). *The Routledge Handbook of Research Methods for Social-Ecological Systems*. Routledge.
Brint, S. (2002). The rise of the "practical arts." In S. Brint (Ed.), *The future of the city of intellect* (pp. 231–259). Stanford University Press.
Carr, D. (2014). Professionalism, profession and professional conduct: Towards a basic logical and ethical geography. In S. Billett, C. Harteis, & H. Gruber (Eds.), *International handbook of research in professional and practice-based learning* (pp. 5–27). Springer.
Colwell, R. (2012). Pride and professionalism in music education. In G. E. McPherson, & G. F. Welch (Eds.), *The Oxford handbook of music education* (Vol. 2, pp. 607–611). Oxford University Press.
Cribb, A., & Gewirtz, S. (2015). *Professionalism*. Polity.
Dyball, R., & Newell, B. (2015). *Understanding human ecology: a systems approach to sustainability*. Routledge.
Dzur, A. W. (2008). *Democratic professionalism: Citizen participation and the reconstruction of professional ethics, identity, and practice*. The Pennsylvania State University Press.
Esposito, E. (2017). An ecology of differences: Communication, the web, and the question of borders. In E. Hörl & J. Burton (Eds.), *General ecology: The new ecological paradigm* (pp. 285–302). Bloomsbury Academic.
Evetts, J. (2014). The concept of professionalism: Professional work, professional practice and learning. In S. Billett, C. Harteis, & H. Gruber (Eds.), *International handbook of research in professional and practice-based learning* (pp. 29–56). Springer International Handbooks of Education.
Fenwick, T. (2016). *Professional responsibility and professionalism*. Routledge.
Folke, C. (2006). Resilience: The emergence of a perspective for social–ecological systems analyses. *Global Environmental Change, 16*(3), 253–267.
Gale, T., & Molla, T. (2016). Deliberations on the deliberative professional: Thought–action provocations. In J. Lynch, J. Rowlands, T. Gale, & A. Skourdoumbis (Eds.), *Practice, theory and education: Diffractive readings in professional practice* (pp. 247–262). Routledge.
Gonzales, M. (2020). *Systems thinking for supporting students with special needs and disabilities*. Springer.
Greene, M. (1995). *Releasing the imagination: Essays on education, the arts, and social change*. Jossey-Bass.
Hargreaves, A. (2000). Four ages of professionalism and professional learning. *Teachers and Teaching, 6*(2), 151–182. https://doi.org/10.1080/713698714
Hargreaves, A., & O'Connor, M. T. (2015). *Collaborative professionalism: When teaching together means learning for all*. Corwin.

References

Haycock, J. (2015). Protest music as adult education and learning for social change: A theorisation of a public pedagogy of protest music. *Australian Journal of Adult Learning, 55*(3), 423–442.

Hertzke, A. D. (1998). The theory of moral ecology. *The Review of Politics, 60*(4), 629–660.

Himes, A., & Muraca, B. (2018). Relational values: The key to pluralistic valuation of ecosystem services. *Current Opinion in Environmental Sustainability, 35*, 1–7.

Hunter, M. A., Aprill, A., Hill, A., & Emery, S. (2018). *Education, arts and sustainability: Emerging practice for a changing world*. Springer Briefs for Education.

Ilmola-Sheppard, L., Rautiainen, P., Westerlund, H., Lehikoinen, K., Karttunen, S., Juntunen, M.-L., & Anttila, E. (2021). *ArtsEqual: Equality as the future path for the arts and arts education services*. CERADA. https://urn.fi/URN:ISBN:978-952-353-043-0.

Ison, R. (2017). *Systems practice: How to act in situations of uncertainty and complexity in a climate-change world* (2nd ed.). Springer.

Koivisto, T. (2022). *The (un)settled space of healthcare musicians: Hybrid music professionalism in the Finnish healthcare system* [Doctoral dissertation]. Studia musica 89. Sibelius Academy. University of the Arts Helsinki. https://urn.fi/URN:ISBN:978-952-329-264-2.

Long, J. C., Cunningham, F. C., & Braithwaite, J. (2013). Bridges, brokers and boundary spanners in collaborative networks: A systematic review. *BMC Health Services Research, 13*, Article 158. https://doi.org/10.1186/1472-6963-13-158.

Midgley, J. (1981). *Professional imperialism: Social work in the third world*. Heinemann.

Morton, T. (2018). *Being ecological*. The MIT Press.

Plaisance, P. L. (2021). The concept of moral ecology in media sociology research. *Communication Theory, 31*, 422–441. https://doi.org/10.1093/ct/qtz022

Sachs, J. (2003). *The activist teaching profession*. Open University Press.

Sachs, J. (2016). Teacher professionalism: Why are we still talking about it? *Teachers and Teaching, 22*(4), 413–425. https://doi.org/10.1080/13540602.2015.1082732

Sandlin, J. A. (2009). Understanding, mapping, and exploring the terrain of public pedagogy. In J. A. Sandlin, B. D. Schultz, & J. Burdick (Eds.), *Handbook of public pedagogy: Education and learning beyond schooling* (pp. 1–6). Routledge.

Sandlin, J. A., Burdick, J., & Rich, E. (2017). Problematizing public engagement within public pedagogy research and practice. *Discourse: Studies in the Cultural Politics of Education, 38*(6), 823–835.

Saxena, A. K., Chatti, D., Overstreet, K., & Dove, M. R. (2018). From moral ecology to diverse ontologies: Relational values in human ecological research, past and present. *Current Opinion on Environmental Sustainability, 35*, 54–60.

Schön, D. (1983/2016). *The reflective practitioner: How professionals think in action*. Routledge.

Senge, P., Smith, B., Kruschwitz, N., Laur, J., & Schley, S. (2008). *The necessary revolution: How individuals and organizations are working together to create a sustainable world*. New Doubleday.

Sennett, R. (2008). *The craftsman*. Yale University Press.

Solbrekke, T. D., & Sugrue, C. (2011). Professional responsibility—Back to the future. In C. Sugrue & T. D. Solbrekke (Eds.), *Professional responsibility: New horizons of praxis* (pp. 11–28). Routledge.

Spillman, L., & Brophy, S. A. (2018). Professionalism as a cultural form: Knowledge, craft, and moral agency. *Journal of Professions and Organization, 5*(2), 155–166. https://doi.org/10.1093/jpo/joy007

Sullivan, W. M. (2005). *Work and integrity: The crisis and promise of professionalism in America* (2nd ed.). Jossey-Bass.

Väkevä, L., Westerlund, H., & Ilmola-Sheppard, L. (2017). Social innovations in music education: Creating institutional resilience for increasing social justice. *Action, Criticism, and Theory for Music Education, 16*(3), 129–147. https://doi.org/10.22176/act16.3.129.

Väkevä, L., Westerlund, H., & Ilmola-Sheppard, L. (2022). Hidden elitism: The meritocratic discourse of free choice in Finnish music education system. *Music Education Research, 24*(4), 417–429. https://doi.org/10.1080/14613808.2022.2074384

Westerlund, H. (2008). Justifying music education. A view from the here-and-now value experience. *Philosophy of Music Education Review, 16*(1), 79–95.

Westerlund, H. (2019). The return of moral questions: Expanding social epistemology in music education. *Music Education Research, 21*(5), 503–516. https://doi.org/10.1080/14613808.2019.1665006

Westerlund, H., Karttunen, S., Lehikoinen, K., Laes, T., Väkevä, L., & Anttila, E. (2021). Expanding professional responsibility in arts education: Social innovations paving the way for systems reflexivity. *International Journal of Education and the Arts, 22*(8), https://doi.org/10.26209/ijea22n8.

Chapter 4
Using Moral Imagination

Abstract This chapter presents a narrative of music educator and entrepreneur Tuulikki Laes and her work in community-based projects in Finland. Laes' work is theorised through Liz Lerman's metaphor of "hiking the horizontal" through the social ecologies in which we live and work to effect ecological systems transformation in the provision of music engagement and education for previously disenfranchised social groups, including older adults and those living with disabilities. She exercises a "moral imagination" to promote intergenerational music engagement and education that works across boundaries of the field and societal expectations and norms. Like Lerman, she asks artists to consider paradoxes, seek opposites, and reject dichotomies that prevent equitable access and participation in music and instead embrace multiple forms of knowledge and practice.

Musical Beginnings

Tuulikki Laes' website describes her as "a researcher and academic entrepreneur who believes in the potential of music and the arts to tackle major challenges in our society." She is also the founder and CEO of RockHubs, "a social enterprise that helps to build accessible, intergenerational music communities in urban environments." Where did this commitment to effecting social change through music action emerge? Tuulikki begins her account of her engagement with music stating that *"it goes back to my childhood. My father was an amateur trumpet player, my aunt was a music teacher and music critic, and my grandmother was an opera singer. So, I had music in my life. One of my first memories is that my dad used to play jazz records to me when I was very little. I kind of grew up with music in a very natural way."* As with many Finnish children, at age 6 Tuulikki sat the entrance exam for music school and started piano lessons. She explains *"my becoming a musician was very path-dependent … I went through the music school system with all the level exams and music theory and followed this very … stereotypical musical education we have here in Finland."*

Music as a "Lifestyle"

While Tuulikki describes her coming to music as "path-dependent" for her father it was not so organised. "*He was 14 when he started playing the trumpet, so he started quite late. And I think his musicianship has always been a bit more … how to say, passion-driven. Although he has made his professional career elsewhere, he's always had these brass bands that he has put together with his friends. They have been very passionate, very engaged in their hobby that has almost become another profession. They have made albums together, they've had a lot of gigs together. My mother always talks about how their marriage was basically defined by these brass bands and their schedules and it was part of their lifestyle, that these other band players and their families were close to our family and we went on trips together. I almost considered him more of a musician than I am because it was a passionate thing for him to have this other identity as a trumpet player based on social life and community.*"

Tuulikki continues, "*I never really considered whether or not I should pursue this musical career, it was more like a natural thing for me. There's always been music, music was something that I felt was kind of easy for me. It was comfortable. I like to go to piano lessons. I followed the plan that was given to me by the music school system.*"

Stepping off the Path

While the School system provided Tuulikki with a plan, it also offered some divergent paths through a music education. She continues, "*around 13, I signed up for this class by another piano teacher who wasn't my own teacher who taught so-called rhythm music, like popular music and jazz. His teaching style was very different and I wanted to learn something different, not only classical piano. And I think it was the first or second lesson when he just put a microphone in front of me and said 'Okay, now you can sing and play the piano.' I'm like, 'No, I can't, I can't do that at the same time.' But he was very decisive. He led me into a whole new world. We went through these older folk music classics like Carole King and Linda Ronstadt, and I just basically learned that style. It opened a whole new world to me, a world of jazz and popular music and improvisation and singing. I started singing in a couple of bands, we had a jazz trio, and then we had another band where I played the piano and sang. We performed in the music school concerts and then just escalated from there.*"

From these beginnings Tuulikki's teacher became the band's manager, organising "*gigs in art exhibition openings or in nightclubs.*" She explains, "*I expanded my musicianship as a popular rock band musician, taking up bass guitar*" and performing more as a jazz, popular and rock musician. She describes this period as a teenager as "*kind of closing a circle*" as her "*first musical memories relate to jazz music and that's something that I've always shared with my dad, the love for jazz music.*"

This "*closing a circle*" culminated with opportunities to perform as the vocalist in her hometown's Big Band, alongside her father, including performances of Duke Ellington's jazz mass.

Studying Music Education

It was through a casual conversation with her aunt that Tuulikki heard that the Sibelius Academy had a music education department. She had dismissed the idea of studying music performance thinking "*I'm not good enough. But my aunt said, 'Yeah, you can apply,' you don't have to perform in the same way as you have to, if you want to apply to the performing department. I was also interested in actually studying academic subjects ... and I realised that this would be a way for me to combine these two; music and academic subjects. I think it just felt a natural choice for me because music is something that's kind of like familiar to me. I don't know if I really had the ambition of becoming a top musician ... it was more like, 'What else could I do, except something that relates to music?' It ... felt so natural.*"

Tuulikki's journey to the formal study of music education had prepared her well. Her pedagogical experiences had encompassed formal music teaching through classical piano, non-formal music teaching through the school-based rock and popular music bands, and informal music teaching through the community work with her father's band. In these settings she had encountered a range of different pedagogical styles and methods including "*group pedagogical work and improvisation.*" Reflecting on these experiences, Tuulikki expands on the affordances of small-town life where "*it was kind of natural*" to undertake many roles rather than specialising in one. For example, her "*music theory teacher in music school was also a part-time composer. He composed a musical that was related to the history of our hometown ... When that performance was put together, they needed a bass player and I had been playing bass maybe for a year and I thought, 'Okay, well, I can come and play the bass.' When I think about it, the classical piano career was the core of my musicianship ... that's the kind of real musical studies that I have to take in order to, for example, get into Sibelius Academy. But then everything else I did around that was something that I didn't stress about so much ... I considered that more as a tool for me to be able to do all these productions and band performances. I had a different relationship with classical piano than with everything else that I did ... I perhaps settled with having this kind of generalist musicianship identity because I knew that I will never become a concert pianist. [Classical piano] was just something that I had to do because it was part of the music school system.*"

Recognising One's Strengths

Tuulikki describes her years of study at the Sibelius Academy as a period during which she started to identify her own strengths and potential for the first time. She explains, "*As part of our studies, we did a lot of group activities, playing in bands and developing interaction skills that we were studying through improvisation techniques. And then, of course, we had our first teaching practice already at the end of the first year; we went to primary school to give some music lessons and I think it was, again, really easy and natural for me. I had also been teaching in a music playschool; I had 5 and 6-year-olds and I also did a small musical with them, we performed it to the parents. So, it's something that probably comes naturally for me that I can handle groups, I can be the inspirer and encourage people to engage in this kind of group activities. But I never really thought about it, it was just something natural. During this first year of studying, I became more aware of how my course mates have different skills and then through that, I started identifying what my strengths are.*"

She summarises these strengths as a capacity to "*handle groups … and a type of leadership that is not intimidating but … makes the group dynamic work, taking into consideration different personalities in the group, making sure that everyone is feeling comfortable. It's kind of this skill to read other people also in nonverbal ways.*"

Tuulikki has drawn on this "*situational awareness*" from the time of her early work with Resonaari, a music school for people with special educational needs. She continues, "*that was the time when I completely gave up my earlier ideas of the norms and structures of music education that I learned through my path-dependent past. When I entered the music school, I realised that I can't actually make use of anything that I have learned as a musician before. I just have to make use of these skills that I have, that I can spontaneously throw myself into interaction with other people. I still think that it was the students there who actually taught me to become what I am today …. We didn't talk about music theory or anything that I used to think are the components of music education that we have to learn. There I was working with students who didn't have the same physical or verbal skills, so we had to find other ways to help them to make music. And it was the trust that really enabled me to throw myself into that learning while I was teaching; trust from the leaders of Resonaari.*"

Learning to Trust Relationality

Working at Resonaari was a formative experience for Tuulikki. She tells the story of her first piano lesson with a teenage boy with Downs syndrome. "*When he came in, I was thinking like, 'Okay, let's have a chat. Hi, I'm Tuulikki, I'm your new piano teacher. What's your name?' And he was just sitting there, quiet for a long time and I'm asking, 'So, what's your name?' He said 'I don't remember right now.' And then*

I figured that maybe chatting is not the way to start musically, maybe he's confused because I'm trying to interview him. So, I started playing and then we played four hands and I noticed that he doesn't remember his name but he knows that red is C, and now if I play red, he starts red. I showed him three colours and I told him 'You can use these colours and then we improvise together.' I realised then, this is what I actually have to do, I have to just make music with these students. I have to find the core of what music-making is about, instead of problematising it or academising it."

Tuulikki expands on her experiences working at Resonaari, describing a student group with many different emotional and social problems who "*carried a lot of burdens with them.*" She stresses that her role was to teach music, not to be a therapist taking "*that burden from them because that's not your job. Your job is to teach music and that's how you can help. I created some type of professional way to handle this because there were very hard stories behind many students. I had a 6-year-old who had a brain tumour and who didn't have a long time to live. And somehow, we ended up having really nice piano lessons, I just focused on making music with him. I knew that his mom was really tired—I could see that from her face—and I think that during that piano lesson she could have 30 minutes of just unwinding outside the room. But now when I think about it, I don't actually know how I could bear all of those stories. I think it was just this kind of professional approach that I took: that we make music and that's enough.*"

Generating Systems Transformation

A Call to Research

Tuulikki's early life was surrounded by models of music engagement. She was also surrounded by models of academic engagement. Both of her parents had academic careers in education sciences, and had completed degrees; her mother in education and her father in psychology. "*I have been surrounded by books and academic concepts ... that's also part of my childhood,*" she recalls. In another "*natural turn*" she relates that she had already decided "*to engage in doctoral studies because I wanted to study more and understand more and have more theoretical and conceptual skills rather than just teaching music.*" Part of the impetus to undertake doctoral studies came from a growing frustration at having "*to defend, alongside with my colleagues, our students' (at Resonaari) right to study music and be serious music learners and creating pathways for them to actually become musicians. Because there were many situations where I noticed that they were considered somehow lesser as music students. Their caregivers, parents, and outsiders were defining what we do as music therapy or some type of recreational music, you know, club that doesn't really have any goal. It's just nice that they can do a little bit of music there and that was against the philosophy that Resonaari had.*" Tuulikki's studies were underpinned by a view that the students at Resonaari had "*musical pathways and learning and *goals .... Beyond a music therapy view ... they can actually be music learners with ambitious goals. And that led me to my doctoral studies."*

A Researcher in Her Own Terms: An Academic Entrepreneur

Tuulikki describes her current professional role as *"first and foremost, a researcher but a researcher in my own terms ... I don't see just one path of being a scholar for me. I actually resist a little bit ... having a certain type of scholarly path. Instead, I like to think about myself as an academic entrepreneur who makes use of a researcher's skills and scholarly background to actually make an impact .... My researcher identity is more impact-driven than publication-driven. I know that I have to make publications in order to get funding and it's also important to participate in academic conversation but that's not the only goal that I see for myself as a scholar. So, that's why I have established other ways to fulfil my ambitions and goals as a researcher."*

Tuulikki attributes her focus on impact-driven research to her time as a teacher in Resonaari, an experience which changed her view on *"who music education belongs to, who is entitled to be a musician."* That experience prompted her to question her own pathways through music learning, to recognise her *"own entitled and privileged position ... this path-dependency that I had."* She describes a process of detaching herself from that position in order to *"recreate my skin as a pedagogue and musician."* She continues, *"I want to make an impact that helps people to have their own self-defined relationship with music, to be musicians in their own terms, to be able to have an identity of something else than what is pre-assigned for them from outside ... whose musical paths might be restricted .... And I'm not only talking about people with disabilities, I'm thinking in plural terms about human beings."*

Tuulikki describes the ways in which her engagement in public discussions via blog posts, media interviews, and panel discussions is a crucial element of her impact-driven research work. Recently, she was invited to give a talk at an event organised by the Finnish Institute in Japan *"on creative ageing and approaches to cultural participation in later life."* Her research collaboration with the largest Finnish pension insurance company Ilmarinen has generated extensive media attention about older adults' participation in the arts and culture in Finland. Being involved in a major research project ArtsEqual (2015–2021) was particularly significant in the formation of Tuulikki's researcher identity. This *"nationally and internationally unique project was funded by the Strategic Research Council of the Academy of Finland which indicated that the research conducted within the programmes must be solution-oriented and the research needs are determined by the Finnish Government."* Tuulikki stresses the importance of active collaboration between those who produce research knowledge and those who use it requiring *"researchers [who] are trained for making policy briefs and participating actively in the political discussions and decision-making processes."* Tuulikki's key role in the project required her to attend several events aimed at politicians and decision-makers, at a time when she was able to *"gain more experience about how research can actually make an impact and how to communicate research in a way that is relevant to people beyond academia.*" She has been able to use this knowledge later in her work as an academic and an entrepreneur. As she explains it, "*I don't see music and art and community work or social justice work or activism somehow juxtaposing with entrepreneurship, I think that's the future. I am building my own professional career and landscape in my own terms. I'm an academic entrepreneur and I follow my values and I know that is the right thing to do.*"

Living Relational Values

Tuulikki's ambitions for all to have a self-defined relationship with music are underpinned by a firm commitment to the values of "*Justice, social justice, and solidarity, equity, and kindness.*" As an entrepreneur Tuulikki has co-founded (with Arash Sammander) a social enterprise company RockHubs as a means of broadening her professional landscape and capacity for impact. The company mission is to:

> help people create a sense of community and have a positive impact on individual identity and well-being through accessible music learning environments. This will foster openness, understanding, and solidarity between neighbours regardless of their age, ability, and background. (https://www.rockhubs.com/about)

With complementary skills in music education, working with differently abled people, and research (Tuulikki) and expertise in design and company start-ups (Arash) the company's mission is to "*build accessible music learning communities for everyone regardless of their age and ability and help to bring people together through music-making and performance.*" Tuulikki noted that many people had a complex relationship with music, where "*music represents something that has created anxiety and shame ... I started thinking, 'How can we help people to create a lifelong positive relationship with music and music-making?' 'Can we fix it in old age?' A Finnish psychotherapist Ben Furman wrote a best seller a few decades ago called,* It's Never too Late to Have a Happy Childhood *and I was thinking, 'Well, it's never too late to have a happy musical relationship either.'* " RockHubs' mission is implemented through opportunities for intergenerational groups to create, play, and inspire. As noted on the RockHubs' website, these incorporate a belief that everyone can be a musician, playing from day one, that creating together "helps grow trust and solidarity" and crucially that it is through public performance that music engagement can "help to create new identities, change stereotypes, and inspire others" (https://www.rockhubs.com).

Taking Advantage of Serendipitous Moments

Tuulikki describes a series of seemingly serendipitous moments that launched Rock-Hubs from idea to existence. *"When we started, I think it was a lot of coincidences that helped us to take off so fast because we didn't have any money. We thought we can't start up a company by taking money in person because that's not the kind of equity-based work that we wanted to do, so we have to find other ways to make this happen. We went to this innovation competition organised by the University of Helsinki. We submitted our idea there but it didn't go through, but we still decided to go and listen to what kind of ideas got the funding. And in the event, there was a Finnish politician who was in the ministry at that time, and I really admire him. He's a very interesting politician and he was talking in the event. And after the event, I quickly pitched the idea of RockHubs saying 'Hi, we also participated in the competition with our idea of RockHubs that creates intergenerational music learning spaces for people.' And he immediately picked up and said, 'That sounds interesting. My friend is a CEO of an affordable housing company and they are just building an intergenerational residential block in Helsinki. You should contact him, I'll put you together in an email.' And then it happened. The same night, after midnight, he had written an email and introduced us. And then I met this CEO next week and he really liked the idea and so we got the space and started our activity there. That's how it took off without needing to have any funding because we got the space there for free."*

Working Across Boundaries

Tuulikki's work is shaped through her history and experience of working across the boundaries of disciplines including the boundaries between music education and music therapy, between music education and community music, between academia and public-facing approaches to inquiry and understanding. Her focus on intergenerational work might be viewed as another example of working across boundaries to facilitate collaborations in spaces that have not previously existed. She describes herself as having *"one foot in the scholarly field and my other foot somewhere else. Because of this I feel we are able to do things outside of definitions, which I find really empowering. I can give you an example of that. Last year, when we were still working in this generation's block, we had a few bands running there on a regular basis. A couple of families approached us with small children, asking 'Is it possible that we could also form a band? Could we play together with our children?' One of the Resonaari teachers who was helping us by leading or facilitating these bands said, 'It's a bit difficult, you know, because we need a different pedagogical approach with children and adults, so how to actually build this? This pedagogical situation is not so straightforward, so I don't think it works.' I started thinking 'Why not? Why not just give it a try? We don't have to follow these rules anymore, we can do whatever* we want.' We started as a pilot with one family, mom, stepdad, a 5-year-old, and a 12-year-old. We recorded that pilot and documented what's going on and tested different pedagogical ideas. And this year, we got a grant from a foundation and we started a family band project with five families. So, now we have five bands and it actually works really well. Because of the COVID situation, these families already are together, so it's easier in those terms. Our youngest participant is 5 [years old] and it's working really well. So, I don't know what the fuss is about, having separate pedagogies with children and adults."

In addition to receiving grants for the family band project, RockHubs has also been awarded with funding from various foundations to carry out music projects among marginalised youth and language minorities, as well as a project at a nursing home for people with dementia and brain injuries. As part of these projects, RockHubs provides mentor training for the caregivers to be able to continue the music activity independently after the project funding is over. The mentor training was piloted in Canada in 2019 as part of a research–workshop collaboration between RockHubs and the Universitée Laval in Quebec and is particularly designed for supporting people with non-professional music backgrounds to facilitate music activity in contexts such as elderly care and disability services.

In 2021, RockHubs received funding for 2 years from the City of Helsinki to launch a "block band" project in four different locations in Helsinki. Three locations are community apartment buildings for senior citizens and one location is a cultural centre accommodating culture and well-being services for newly arrived immigrants and refugees. Tulikki describes the main goal of this project "*to provide ongoing group music activities in a rock band setting and in this way strengthen the research-based service model developed at RockHubs.*" With the help of joint live-stream and in-person events, the project aims to create a functioning and long-term cultural network between the participants at all four locations. In this way Tuulikki continues, "*the project aims to promote cultural inclusion and to develop a local social ecology where communities and diverse groups of people can together enrich their living environment and support each other through shared musical experiences.*" As part of the project, Tuulikki has curated a one-day outdoor festival, "The Source of Life" that celebrates intergenerational solidarity and the lifelong power of the arts by presenting multiple senior performers (including the RockHubs' bands) in collaboration with performers from the fields of circus, dance, and theatre. A further event will be launched as a result of the RockHubs partnerships under the title "Senior Rock 2022" hosted by the Cultural Center Cable Factory owned by the City of Helsinki.

Changing Perceptions, Beliefs, and Key Goals

Tuulikki's work is contributing to changing perceptions of what music education is, and for whom. This work also changes perceptions of social structures and human experiences. For example, the rock bands for older adults are changing "*the perception of ageing for younger audiences.*" Tuulikki lists amongst her beliefs the view that, "*music specifically, can be used for tackling different social problems and challenges in our society. I'm not talking about the instrumentalised ways that some others might be after. I think it's enough that every human being has a potential musical relationship, their own individual relationship with music, and music enables everyone to explore. I want to see ... that we understand the potential of music in tackling different types of situations and challenges that are part of human life and life course.*" This work involves "*dismantling certain myths and norms and structures around musicianship and music education; being critical to the instrumentalising of music, reducing it to some type of medicine that is used, in therapeutic or care contexts nowadays more and more; or being critical to this brainification discourse, that music makes you smarter .... I'm really critical towards that because I think it's enough what is happening.*"

Showing not Telling

Tuulikki describes her approach to music education and music-making as one where "*people simply engage with learning music, practising, and then becoming musicians in their own terms, opening up new insights, facing difficulties, as part of becoming a musician is facing challenges ... overcoming and continuing that work and creating new goals, new ambitions, individual and shared goals in music. I think that's enough. That's the power of music, we don't need brain scans to show what happens in which part of the brain when you play the piano. We don't need statistics on how listening to music lowers the blood pressure or we don't even need to define and categorise and label people in terms of creating these categorised spaces; for example, music for marginalised youth, music for refugees, or music for senior citizens or elderly people. We don't even need those categories. We just need a space where people, irrespective of their skills or abilities or background, come together and there's someone, a facilitator, a teacher who helps them to start learning and practising music and then they perform that music to others. And they can show, 'This is who I am, this is what I can do, this is who we are,' and then that inspires other people to see maybe I can do that as well, or I see that there's different aged people playing together, or there's a family band with two mums, that's cool. That's a simple recipe for how music can change attitudes and create solidarity and construct positive identities among people.*" In these examples, Tuulikki's approach to public pedagogy is foregrounded as one of showing, not telling, of offering different examples of how life is and might be in and through engagement with music.

As we discuss public pedagogy Tuulikki tells the story of Kaisa who, aged 65, took up bass guitar as a member of the "Grannies Band" and became an advocate for music education for older people and Resonaari. Tuulikki recounts "*one of the reasons for that was that it made such a huge difference in her own life .... I invited her to a few of my lectures when I was teaching music education in later adulthood at the Sibelius Academy .... I think she has really inspired me to pursue what I'm trying to build through RockHubs and also as a scholar .... I'm still meeting with her regularly.*

We are having conversations and I'm documenting them and we are planning to write a book together .... She has taught me a lot about ageing in general, about how she thinks about her life course. We have together engaged in these mutual learning experiences. She has given me new insights about the so-called value of music learning, she has been able to connect back to her earlier life history in a generative way. Some of the memories that she has shared with me have included feelings of shame which she has been able to turn into positive narratives. So, it's really a reciprocal interaction and relationship that I have with her and I would consider her as one of my core mentors ... her life experiences and insights, help me reconfigure what I'm actually trying to say in my research and how I interpret the findings and what we are trying to achieve with RockHubs."

Tuulikki's account of her relationship with Kaisa turns notions of public pedagogy upside down, where the role of pedagogue is one which shifts, is reciprocal, and challenges stereotypical notions of who can be the pedagogue. Tuulikki recognises a *"pedagogical element in my work as a core for everything that I do because I aim to offer people different perspectives in the hopes to help them to engage in learning something new about what they think that they already know. So, in that sense, I think that whether it's facilitating groups or lecturing for music teacher students or public speaking or writing an article, I think that's at the core of everything I do ... public pedagogy is not a pedagogical process happening between two individuals inside a structured educational space or situation, but something that happens in a more broad and complex contexts and ways."*

Hiking the Horizontal

Tuulikki Laes describes her desire to live in a non-hierarchical professional "laboratory of doing," hiking the horizontal in Liz Lerman's (2011) words (see Fig. 4.1). For Lerman, "hiking the horizontal" provides "a kind of mnemonic device to dismantle hierarchies that are so embedded in our cultures that we cannot imagine anything else" (p. 292). Lerman asks artists to consider paradoxes, seek opposites, and reject dichotomies that prevent many ideas from coexisting; to look instead for synthesis, and foster learning communities that embrace multiple forms of knowledge and discovery (p. xvi); to understand that not all distinctions need to be "about right and wrong" and rather make "the walls permeable between these distinctions" (p. xvii). This kind of hiking through non-hierarchical social ecologies is shaped by systems reflexivity and imaginative acts in a laboratory of doing. Such a laboratory requires also non-conventional solutions that challenge established institutions, their boundaries, and the vertically functioning professional value hierarchies (the progression within the cognitive ladders of musical practices) and epistemic assumptions that shape music education theory and practice as well as professionalism in music.

For Lerman, the process of hiking the horizontal is a critical response process, "a kind of ritual" that "holds you in relationship to values that would be otherwise hard to keep in the rush of daily living" (Lerman, 2022, p. ix). The process "is a set

Fig. 4.1 "Hiking the Horizontal" in professional life

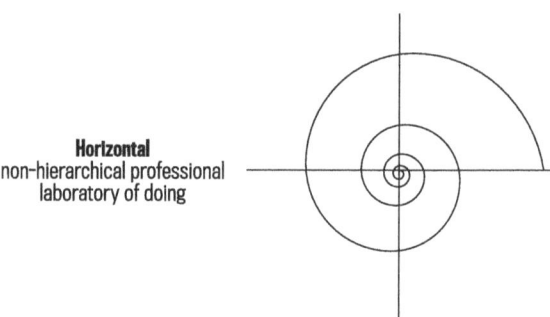

of practices that allows you to live on the horizontal in a world dominated by the hierarchical … it can help you manage ambiguities, hold multiple perspectives, and function where meaning depends on context but where values and ethics still matter" (p. ix). Through the use of dramatic rehearsals and ritualising difference, we can foster moral imagination that engages music education with the wicked problems of the world, at deep levels, in order to generate systems transformation.

Tuulikki's professional life in the laboratory of doing is animated by a socially responsible, activist, and morally imaginative mindset. She is committed to finding and pursuing the public pedagogical possibilities of "grassroots organizations, neighborhood projects, art collectives, and town meetings—spaces that provide a site for compassion, outrage, humor, and action" (Brady, 2006, p. 58). Such socially responsible work seeks to move beyond the "silo mentality" of professional practice, to interrogate the professional system through a process of systems reflexivity, and create alternatives to institutionalised practices. This work has inherent risks and requires courage as professionals challenge the prevailing value hierarchies and epistemic assumptions of social-ecological systems (see Fig. 4.1).

In her doctoral dissertation Tuulikki (Laes, 2017) takes a reflexive stance by calling for an approach in which self-conscious recognition and the relational nature of teaching as well as research are acknowledged: a kind of reflexive critical pedagogy that moves beyond the often- uncompromising positions that characterise critical pedagogy. For her, activism relates to "the production of knowledge, policy, and pedagogical practices through active engagements with, and for, social groups that are marginalised by society" (p. 43). She writes that in her life, "activism refers to the processes of addressing challenging research questions, elaborating methodological considerations, and exercising reflexivity, as well as engaging in personal confrontations with my teaching practice" (p. 43). Activism is thus not simply supporting the status quo of the existing system with a collaborative approach; an activist mindset examines social ecologies in one's environment through imaginative lenses, thinking beyond "what is" to "what could be." She writes:

> Tapping into a situation's possibilities with an appreciation towards *the position of the other evokes deliberation through disrupting action that, according to Dewey, can be considered as "a dramatic rehearsal" of moral imagination* [emphasis added] ... [in which] acknowledging the multiplicity of possibilities for acting in problematic situations requires sound deliberation through moral imagination, rather than following a reductionist view of a single moral etiquette as if there was only one solution available to every situation. (p. 43)

Recognising the "multiplicity of possibilities" inevitably invites complexity. As Tuulikki notes "there is no reason to reduce or dim this complexity; rather, it gives a generative aspect to the regularities and irregularities of educational action and interaction" (p. 44). Complexity provides a different approach to professional responsibility requiring more than inventing new teaching methods "based on a preconceived notion of what is 'good' for the students" (p. 44). Thus, ecological systems transformation requires a different orientation to complexity, an orientation that invites amplification of complex issues, to consider them at the micro, meso, and macro levels as part of framing the problem. These processes require the exercise of the moral imagination within the spheres of action and influence in which we work. Tuulikki's work can be viewed as an exemplification of the ways in which local action can not only make a difference, but also provide exemplars for others to consider. While this work may be viewed as a partial solution only, its practice can provide ways to re-frame problems, to consider and experiment in practice with other possibilities, and engage with processes of active transformation.

References

Brady, J. F. (2006). Public pedagogy and educational leadership: Politically engaged scholarly communities and possibilities for critical engagement. *Journal of Curriculum and Pedagogy, 3*(1), 57–60.

Laes, T. (2017). *The (im)possibility of inclusion. Reimagining the potentials of democratic inclusion in and through activist music education* [Doctoral dissertation], The Sibelius Academy of the University of the Arts: Studia Musica 72]. https://taju.uniarts.fi/bitstream/handle/10024/6606/nbnfife201705096359.pdf.

Lerman, L. (2011). *Hiking the horizontal: Field notes from a choreographer*. Wesleyan University Press.

Lerman, L. (2022). Preface: The critical response process. In L. Lerman, & J. Borstel (Eds.), *Critique is creative: The critical response process in theory and action* (p. xi). Wesleyan University Press.

Chapter 5
Ecopolitical Systems Reflexivity in Practice

Abstract This chapter presents a narrative of musician, music educator, and festival organiser Riju Tuladhar and his work in multiple arts projects in Nepal. The chapter outlines his work to sustain the cultural heritage of Nepal while at the same time effecting systems transformation in cultural practices that prevent inclusion, participation, and sharing for all. His work is grounded in the diverse and multiple musical ecologies of the country and seeks to build career opportunities and learning pathways for the next generation of musicians and music educators. The narrative illustrates the ways in which ecological values and practices underpin collaboration and cooperation that promotes systems transformation. Riju's work in public pedagogy merges a humanitarian perspective with a sustainable cultural practice that builds futures for Nepali music.

Professional Life of a Musician in a Land of Festivals

> Nepal today is at a stage where most of the folk and traditional musics, which are an important part of Nepal's intangible cultural heritage, are alive but fading. The country's immense ethnic diversity encompasses diverse languages, customs, festivals and musics. Its tremendous geographic diversity, rising from just a few meters above sea level to the highest point on earth is home to more than 120 ethnic groups, including Sherpas, Tamangs, and Rais in the mountains, Gurungs and Newars in the hills and valleys, and Tharus and Maithalis in the plains of the Terai. The numerous ethnic groups have long lived in distinct communities divided by caste and ethnicity, with each community having its own specific music for performing religious rituals. (Kansakar & Tuladhar, 2020, p. 87)

So writes Riju Tuladhar of his home country Nepal. The intangible cultural heritage of the folk and traditional musics of Nepal is at risk due to multiple challenges posed by the unique environment of Nepal, the significant natural disasters experienced in recent years, and the ongoing adjustments and accommodations of minority world cultural, economic, and social practices and values. Riju's story begins with a reflection on the diverse musical ecosystems in which his music learning occurred.

Music Learning in Diverse Musical Ecosystems

Despite a childhood spent in *"hostels and boarding schools"* for Riju Tuladhar *"home is in the heart of Kathmandu where there is lots of hustle and bustle."* As a Newar, one of the 120 ethnic groups of Kathmandu, Riju's early life was marked by *"lots of festivals ... not the festivals like European festivals, but traditional festivals where every festival has their own music, and own food, own smell, own colours. Whenever I came back home for the holidays ... it was festival. There would always be some music going around on the street or in front of the temple."*

Riju's music experience at home in the Kathmandu Valley focused on the traditional sacred and secular musics of ethnic groups, such as his near and far family members' practice of Dapha (a Nepalese sacred music tradition). Learning these musics was largely an informal affair for Riju as up to the age of 17 he changed school 10 times, spending only a month a year at home during the Dashain time. As he relates, *"Dashain is the longest festival, so that's the longest holiday .... And the training season [for the musicians] is not in the Dashain time, we have to be ready by Dashain time."* He describes his early learning as participation in these festivals by *"playing a little bit of drums, whatever it was in the ethnic group ... so just getting my hands whenever I got my chance to, also ... it was not very open for the other communities to come in and play. I'd need to have months-long lessons to join in. So, we were not quite welcome, but, yeah, whenever we got a chance, we liked to just go around and played a bit."*

At the boarding schools Riju encountered western music through *"some concerts by our seniors ... extracurricular activities ... and club."* It was not long before he started a band *"Kind of rock, punk, blues,"* and released an album which *"topped a chart for more than 6 months in Nepal, and we began touring around Nepal."* Despite that early success as a professional popular musician Riju and his band *"decided to disband because we wanted to learn more about music rather than just school rock, and in that course of trying to learn more, I was actually pulled in to teach at the Nepal Music Center (NMC). And, actually, teaching helped me learn more about music. It paved the way for me."*

Professional Learning Through Teaching

Riju describes his early teaching work at the NMC as a process of *"digging"* into himself, of asking questions: *"What shall I teach? What do I need to teach? And how do I learn, first of all, myself, to teach these things? ... I started learning general, basic music theory ... the ways of learning and teaching came to me by the way that I needed to dig into myself to learn to teach what I want."* Working from textbooks borrowed from friends, the internet, and his lifelong practice of playing by ear, Riju built up his music knowledge and skills and those of his students. He began to think further on what he could offer as a teacher given that his students could learn any song they wanted by ear. *"And then later on I was figuring, okay, what is it that I really want to teach? Are we just, like, teaching that?"* In tackling these questions Riju began to develop his own Nepali music style, his *"own ethnic music. I wanted to write, and I wrote the rhythmic patterns of how it is being sung, and so these things actually helped me to use those particular lines into composing my own music. And I thought that might be brilliant for the senior students ... because other things they could actually find on the internet already ... they were coming up with new phrasings in their ... bass or, in their ethnic drums because these are not available on the internet."*

Riju emphasises the role of communities in his life and work, from his ensemble band in the NMC where he *"used to teach bass and ensemble"* to the communities of the Kathmandu Valley where he took his ensemble, not only to perform but also to *"learn a bit about where actually the community music was going."*

Realising One's Place in Society

Musician for Musician Project: Combining Humanitarian Work with Cultural Sustainability

The 2015 earthquake in Nepal which devastated Kathmandu was a catalyst for Riju to become *"massively, massively involved in the community, helping out communities with relief funds ... building a health post in the Himalayas. I got a lot more entangled, and also very much attached with different communities."* In the time immediately following the earthquake *"there was no role for music, because it was just, like, a really hard time. It was all about whether people ... get to eat and have maybe a tarpaulin shelter, whether they would survive ... after the harsh winter."* As conditions began to settle Riju and his music colleagues noticed that *"lots of music houses, community music houses had collapsed, and lots of community music instruments had disappeared, and the community music teachers, who are very rare ... old ethnic music practitioners, there was maybe one teacher left in one community."*

Riju continues *"For almost a year, villages around Nepal, Kathmandu, and other places hadn't played community music because of the earthquake. And one year after losing the repertoire is big, ... if there is only one teacher and ... there is one year of not playing music and no disciples, no students, then we saw a very big chance of getting our music lost."* Compelled by the prospect of losing even more through the long-term effects of the earthquake Riju began a project, Musicians for Musicians. The aim of the project was to play *"our traditional festival musics."* What had begun as a humanitarian project in the community had become a project in cultural sustainability.

Becoming a Music Education Researcher in Social Change

Riju's journey as a researcher began in 2013 when the Nepal Music Center invited Finnish Music Educators from the Sibelius Academy to work in Kathmandu to help build a music teacher education programme. He describes these initial motivations as developing and changing over the years as *"The Nepal Music Center's top-down plan for building music teacher education together with the Finnish music teacher educators became something else, I think something better, in the co-construction process."* As the two teams worked together, the twin challenges of retaining rock–pop students in the programme—when they could obtain paying gigs around town—and variable levels of knowledge and skill in the student body were highlighted. The solution was to co-create *"a new local lower-level curriculum that could lay the ground for further studies in music education at university level in Nepal."* This project became part of the Academy of Finland funded project Global Visions. Riju identifies a second important change: *"We were offered a possibility to complete our Teacher's Pedagogical Studies through the Sibelius Academy so that our collaborative developmental work could be integrated as part of these formal studies. There are no universities in Nepal that offer such studies and again for free. This was such a motivation for us who were volunteering our time to co-create the curriculum."* Through this programme of study Riju and his colleagues (Kansakar & Tuladhar, 2020) identified a number of distinctive features that shaped Nepali music and music education. As Riju summarises:

"First, Nepal has vibrant cultural practices and music plays a central role in these practices. Through the years, music has been handed down from generation to generation within the community, and music that belongs to one ethnicity has traditionally not been learned or taught to or by any other group. However, with the rise of digital and social media, information has become more accessible and the new generation of musicians have started to see ethnic musics as a common heritage, not limited to specific ethnic groups but as the 'music of Nepal.' New hybrid musical forms were also a way to get round the stigma that often was related to traditional musics and traditional musical instruments, as musicians most often rank low in the caste system.

Second, we identified that the rapid increase in flows of people and media brought on by globalisation following the opening of Nepal's border in the 1950s has greatly impacted music. The influx of hippies in the 1960s, for example, brought, among other things, guitars, violins, cassette tapes, and tape recorders. Since then, the global influence of different types of media on Nepali youths has become huge, drawing them more towards learning and performing western and Bollywood musics, rather than the folk and traditional musics of their local communities. Pop and rock bands started experimenting with folk tunes and instruments creating sub-genres like Lok (folk) pop and folk rock. Curiosity for Nepali music within the aspiring musicians–student[s] has been growing ever since but we also saw that the diversity is a challenging task for the music educators–teaching artists. There was no textbook on what or how to teach in such a new, emerging, diverse environment.

Third, we identified that the political situation has severely affected Nepali music education. Following the end of the civil war in 2006, Nepal began a slow and painful transition towards a republic. This transition has taken much energy from both the government and civil society, leaving the cultural production of musics entirely to the market. As a result, only a few genres of music gain mainstream attention, leading to the decline of many other musics, like traditional and community musics which were lacking professional career opportunities. We had to reflect what kind of responsibility this situation lays on music teachers, particularly if we are to see the immense musical diversity of Nepal as an asset.

Fourth, we noticed that there is a lack of research, documentation, and archiving of Nepali musics, with the limited existing documentation not being publicly available. For example, there are currently no publicly shared archives that house music research taking place in the country. This makes it difficult to access such materials for young musicians, students, intellectuals, and academics."

This analysis of Nepali music and music education systems informed the development of a research agenda. Riju describes this: "*As one of the outcomes from our reflections and understandings, we decided that field trips to villages, making interviews, and documenting the traditional musicians could be included within the curriculum we planned. At the same time this brought in the principle of 'learning in the context' and informal learning aspects within the programme. This also highlighted the advantages of collaborative learning methods. While doing the pedagogical studies, we had already experienced how collaborative studies was a helpful approach for ourselves. Hence, we weaved in the collaborative studies and informal learning that involved a diverse set of social realities within the framework of curriculum.*"

Riju's investigations into the traditional musics of these villages introduced him to "new" traditions such as the work of Nhuchhe Bahadur Dangol who established an "*all housewife choir*" to provide music-making opportunities for women and girls. Intrigued by this choir Riju began a research project that investigated "*What is the social capital the [women] gain? What were the outcomes for the women from participating in the choir?*" His findings noted changes "*from within the family*" where "*after so many years ... the males were starting to support by at least doing a little bit of housework, and then coming into their practice and preparing tea and everything for the women. And so, family support started to come up.*" Two of the women choir participants "*were standing ... for and against the political system ... into the time of elections, they became more active in political or social change. And they got that voice, they were able to come up with that voice ... by playing music in the community themselves. So, there was a social change, there was a change in domestic support, and there was also a change in their psychological selves.*"

Initiating Echoes in the Valley: A Festival for Systems Transformation

The initial investigations into the traditional Nepalese musics outlined above prompted Riju and colleagues to consider the ways in which these musics might be practised, maintained, and drawn upon as a tool for transformation in Nepal's socio-cultural landscapes. Their solution was to found a festival in a land of festivals: Echoes in the Valley. The festival was launched by Riju, Sunit Kansakar, and Bhushan Shilpakar in 2017 and aimed to:

> uncover, revive and make relevant Nepal's intangible heritages and disappearing sounds by showcasing local music, art, and performances of everyday rituals. It transforms small neighbourhoods in Kathmandu valley into grand stages for musical conversations between international and local artists. In addition, the festival offers an array of interactive educational initiatives, music conference, creative workshops, guided neighbourhood walks, and an open museum of communal art and artifacts. (https://www.echoesinthevalley.com/about)

Riju describes a year-long process of development begun in 2016 to bring the community together in Kathmandu. Through *"lots of meetings and compromises"* Riju and colleagues mobilised the community to raise funds to *"reclaim public spaces"* as stages for the festival and invited a number of international academic colleagues from Finland and Sweden to be part of the festival. Those initial conversations and negotiations within Kathmandu brought the community together across the divides of ethnic music practices, of generations, and of caste. As Riju explains it *"the younger generation were with us from the beginning and there was this thing going on … not top-down, but the older people were getting involved in the different activities (with the younger people), like teaching their traditional way of, you know, handicraft and stuff. It wasn't, you know, another activity in the festival."*

That first festival featured a week of international and Nepali musicians learning together in collaborative workshops. Riju explains *"after that workshop, even the youngsters started to show interest in learning traditional music, not only western bass, guitar, keyboards, or drums. They started by things like … whistling drums, whistling music. Slowly they started to come … and say, 'Actually, my grandpa is a good teacher. I didn't realise my grandpa is a, you know, good musician.' So, they slowly started to come in and learn music as well."*

In expanding on the official description of the Echoes of the Valley Festival quoted above, making "relevant" comes to the fore. Riju describes the traditions of Dapha, including practices within *"music groups that do not allow women to touch any music instruments or sing, because they believe … women … because they have menstruation, they are impure, and the music goes straight to God, so women are not allowed."* In an act that he describes as *"public pedagogy,"* Riju and colleagues brought the *"all housewife choir"* to the same stage as the all-male group to perform together in a presentation with a moderator. He describes the instructions given to the moderator: *"We won't talk anything about gender, anything about, these inequalities, but we'll talk about, 'Oh, what's the music the women are playing? What's the music there?' The moderator's going to talk about the differences and the similar case of the music. The idea was just to present them in the middle of the oldest market of Kathmandu, and ... together with the males, only males, and just to irritate them a little, tease them a little, so the males, when they go home, maybe they start thinking, 'Ah, that was nice. Maybe it would be nice if my wife would join, or if my sister would join.'"*.

From that first act of public pedagogy the all-woman group has not only been recognised and celebrated: Their presence has begun to transform the music practices of groups that were previously all-male. Riju notes *"they are already starting to introduce some women into their own groups as well."*

Demonstrating Multiple Possibilities Through Public Pedagogy

Riju describes public pedagogy as a process of posing possibilities: *"not to say this is right or this is wrong, but just to demonstrate what is going on in our community. What could happen? Where could we go? What might be possible? We might already have that in our head, these ... you know, the different possibilities that we ... that might be there that we might be able to create by ourselves, rather than just taking an example of any already existing communities."* Riju expands on this description emphasising that *"operating or coexisting in a community is so different in different communities. For example, the teaching system of Guthi music is totally different ... it's nonacademic in a sense. Especially nonacademic in the sense the education board hasn't identified this way of teaching as music teaching, but Guthi teaching ... music teaching has been there for hundreds of years."*

The Echoes in the Valley festival supports this approach to public pedagogy through providing platforms not only for the performance of many different musics, but also for their distinctive teaching approaches. As Riju explains this, *"bringing in the examples from all these different things together might help us to see how we might be able to do differently, and to ... achieve a better solution, in a way."*

Festival Practices and Values: Collaboration and Cooperation for Ecological Sustainability

From these beginnings the festival has evolved into a presentation pattern of taking place in Kathmandu every second year, with the festival travelling to other villages and regions in the intervening years. From beginnings where musician contributions were voluntary the festival now pays all artists, with all being paid the same amount regardless of experience, reputation, or caste. This has been made possible through Riju gaining financial support, first from colleagues and friends and subsequently from major governmental and NGO agencies. The model eschews sponsorships from commercial enterprises, an approach that has led to sponsorship from cultural agencies such as the Indian Culture Centre, the Swiss Embassy, the British Council, and local Municipalities.

The festival values celebrate local culture and products. "*For example, we don't allow a burger station as a food stall, or a KFC, or Coca-Cola ... but our own local liquors, not imported branded bottled beers, or whiskeys, but we only are open for local community food, not for imported food stations like McDonald's and so on.*" Speaking further of the festival values, Riju describes "*bringing together very different caste groups into the same platform ... brings this issue to the surface. Another ... is the gender thing, with people saying, 'There is no more gender biasness in Nepal.' And actually, there is, because people do not want to look at it or go deep into it. So, we bring those issues not as issues, but, like, just a presentation just to, like, try to push it, and also become a platform for the practitioners, music practitioners and music academics, the young academics, to use this platform as a research ground.*" Riju's commitment to equity, diversity, and social inclusiveness is evidenced in the ways in which the festival operates as a platform for public pedagogy on issues of caste, gender, and the sustaining of local tangible and intangible heritage.

The festival is described as family focused as Riju wanted "*to get the whole of the family in there, so there are ... activities for children groups, there should be activities for adult groups, male or female. ... The living style of Nepal is three generations of a family living in the same house, so there should be activities for grandmas and grandpas. That's why we had lots of children's activities.*" The children's activities encompass book reading in Nepali, Newari, and English, "*with the performances where children are involved. [There are] painting activities, like calligraphy ... and the old games .... You know, because of the videogame, we have lost so many children's games. So, we have children's music workshops. And 2 years back they did it a little bit differently so that they ask the parents to be involved with their children so that they could do something together, and when they go home, they have this activity with the parents at home as well.*"

Riju is alert to the need to document the festival as part of the agenda to sustain this cultural heritage. To this end the festival has a documentary team "*who has been working on the community music development as well, and they are already a part of Echoes as a core team.*"

Strategies Towards Sustainable Futures for the Festival

Riju considers the ways in which Echoes in the Valley can be sustained into the future. He points to the need "*to understand the community at its core,*" a process which will then allow the development of the necessary skills. He emphasises the need to "*not be biased to your own self, and not give up ... to put your ideas, and then have this, like, flexibility to compromise. Not become too rigid, but at the same time, do not, deviate from your goal.*" He suggests that "*If something else rather than your idea can come towards getting the same goal, then it should be welcome. It's more about learning about the community deeply. How it functions. What are the values? And how do we not, like, poke the wrong place, but still try to point out the things that have already been interesting for the community."*

Riju has many roles that intersect in his work to find ways to guarantee the future of the festival. One is as a teacher at the University of Kathmandu where he teaches ensemble. He sees his work as *"kind of intertwined in a way, because Echoes in the Valley becomes a platform for my students to perform as well as do the research work. And whatever they are learning in the field, they are bringing into the classroom. So, it's teaching going on everywhere, as well as learning for me as well as for them. We are bringing lots of different expressions or other influences that we get from the festival, into the classroom, and as well whatever we come up with in the classroom, comes to the festival."*

Amidst all of this he also works with his own band, another example of intertwined practice where *"whatever I learn in the classroom with my students, sometimes I think that in the band, and sometimes what happens in my band, I bring that to the students. In a way, it's very much coexisting. To be honest, it's all a learning process for me. I do not consider myself as a teacher. It's more trying to give a little opening for my younger generation to what I might have seen, and also try to see what they have that I haven't seen."* Riju describes his band as *"a three-piece band, and my other bandmate, he plays sitar and he also plays guitar and is also an ethnic drummer. And another bandmate, he plays mostly all Nepalese and Western drums and all the percussion. And we try to use our experiences as well as the influences that we get from the traditional and festival music of Nepal, and try to use that as a tool for us to make new creations ... analysed fusion, it will always have this Nepali thing in it, but, with whatever knowledge that we have gained from over 20 years of playing music ... we try to come together and reflect ourselves."*

The reciprocity and fusion of experience and practice that Riju outlines is deeply embedded. He traces this back to 2003–2004 and his early work with a colleague with whom he had *"been working on the social issues and community-related matters in the community ground ... we just haven't realised that what we were doing, ... from what perspective we were working."* He continues, *"The music pedagogy and the global research team has given me the vision, made me realise, this is what I have been doing .... So, this is actually what I have been doing to identify myself, my role, ... to find myself."*

Riju also owns and runs a small café as well as teaching 2 days a week in Kathmandu University and 2 days a week at the Nepal Music Center. His commitment to community runs deep, evidenced in another project in which he *"sends volunteers to build a house ... the health post, in the Himalayan ranges, beneath Ganesh Himal. It's the last line of that part of the civilisation. I just finished that."*

Envisioning Amidst Crises

Looking Around and Looking Forward

As for many, "ordinary" life ceased for Riju during the global pandemic. "*The first month of COVID was very intense. There was only one COVID case in Nepal, but we got locked down and everyone was so scared just even to go, like, just out on the street, just outside the house. Everybody was inside the house, there was nothing going on, everything was just frozen. Everything just froze for, like, 2 months. And then slowly, yes, it started to open. The schools are still closed [as of March 2021]. Some schools have started, but I'm still teaching online. And there are not many music performances, but I have started music performance, I've played two gigs in the past 3 weeks, so it's starting. As for festivals, I don't think it's ethical to do, like we used to do before. We are still having discussions on how to do it online or wait until the end of year when the vaccines would already be rolled out to all. The cafés have started to open, but not as before. For the past 3 months I have been focusing more on my café because there is no performance and teaching is ... it's only twice a week for me. At the moment the festival is on hold.*"

Riju describes the shifts in his roles brought on by the strictures of the pandemic, including a focus on "*writing with my friend, Sunit, a band member, as well as my restaurant partner, and festival organiser. We are writing for the Finnish music genre ... people were going into deep depression, especially the musicians, during the pandemic, so we were trying to come up with different videos to support and motivate them towards doing ... you know, daily life [how to] enjoy the daily life.*"

A Vision for the Future of Nepali Music

In thinking about his vision for the future, Riju comments, "*whenever I go and attend in expos, like WOMEX World Music Expo, and whenever I look back into our South East Asia, what we see is only the Indian musicians, and there is a lack of Nepali musicians, and Bhutanese musicians, Myanmar. So, I would love to help the musicians in Nepal to contemporise the folk music. Not only for the sake of contemporising: first of all, to learn something that is different from whatever the world already has, and, in that way, Nepal might have a space to go and present, and if that could be done, then there would be a model for others to follow. And then there is a bigger sense for our youngsters learning this traditional music. I think I would want to take that role rather than just like, okay, let's go and learn music, learn some traditional music, play some folk music. I think that might work in a different way of educating and making people realise …. I would love to see that.*"

Riju's vision also encompasses research: "*I would also, definitely love to see research work done in Nepal. But all the research books and their research papers are locked up with their individual institutions or in the universities. So always the new, younger generations, when they do their research, they always start from scratch. The research world doesn't work like that, you have to have an understanding in the research world of what has already been done. People are always hiding it, thinking that it's their own personal property, and so I want to make it accessible ... because it's public property, it's an intellectual public property. We had been working on making a portal where you can come and find anything. But it is very hard, it will take years for people to open up their resources ... so-called their own private resources. That is one very big goal that I'll be, at least, trying to do. Now, besides festival, yeah, I think these two are for now, my top agendas."*

Riju offers further challenges to music education in his analysis of the ways in which contemporary theory and practice address inclusiveness. He speaks of attending conferences where the keynote *"sounded so much more like bringing in the music education like the European Union rather than to the whole world. And that was kind of, 'Ah, so they are trying to make this one European Union for music education in this particular idea. And how interesting it is also called "inclusiveness".' And we are coming from the place where we have so many ethnic groups, we have coexisted for so long. But what we still need is unification. We need not only to coexist, but to practice unifying diversity."*

As the interview ended, Riju commented on how he views himself as a public pedagogue: *"I question myself a lot and I struggle with that."* Perhaps his pedagogy is best illustrated through his description of the Echoes in the Valley festival as *"a space in which I can work toward the values that I believe in: freedom to work artistically beyond ethnic or gender restrictions and free access for all for learning and developing professional capacities as a musician in Nepal. It has become a space in which I and my colleagues can navigate and envision how the future of Nepali culture might look like."*

Epilogue

Some 2 years after the commencement of the pandemic the Echoes in the Valley festival returned to Nepal. Hosted by Kirtipur on March 26, 2022, the festival animated "existing traditional public spaces such as courtyards, raised platforms and temples ... [to] showcase over 150 Nepali and foreign artists from six different countries across four stages and small nooks and crannies of the old town of Kirtipur" (Press Release March 24, 2022). The 2023 festival occurred in March 2023. Riju is pursuing his vision of a Nepal Music Archive supported by a seed grant from the Goethe Institute of Germany (https://www.nepalmusicarchive.org/home).

Using the Public Nature of Music-Making for Social Transformation

For Riju Tuladhar, sustainability of traditions requires translation and acknowledgement of the changing environment (Kansakar & Tuladhar, 2020), rather than searching for "retrotopias" of the past, as Bauman (2017) called them. Although retrotopias may be the roadmap for some ecologically oriented music education, preservation by turning to past practices with their past hierarchies is not an option for Riju. Rather, he sees that sustaining heritage practices requires adjustments that reflect changing societal practices; for example, through ensuring traditional musicians' livelihoods rather than maintaining their previous unpaid community status. Bauman described the roadmap to Retrotopia as drawing from a "rehabilitation of the tribal model of community" in the belief that it is "the genuine or putative aspects of the past, believed to be successfully tested and unduly abandoned or reck-lessly allowed to erode, that serve as main orientation/reference points" for a better future (p. 9). Instead, Bauman urged for the courage to consider and work with "the not-yet-unborn and so inexistent future" (p. 5). In the same vein, Riju analyses the practices of his own community, and the Nepali music scene in general, and recognises how present and past practices sustain gender exclusion and social structures that are no longer acceptable in contemporary Nepali society. In other words, he has developed systems reflexivity (Westerlund et al., 2021) with ecologically just eyes.

As festivals are the main public form for musicking in Nepali society, Riju chooses this very format to exemplify the possibilities of public pedagogy in the local ecosystems. For Riju, festivals are "not just a music festival, but a statement, an effort at ensuring that Nepali traditional musics as well as the musics of the world are appreciated, in old and new forms, in Nepal and beyond" (Kansakar & Tuladhar, 2020, p. 93). Further, festivals become arenas to enact a public pedagogy in which musical practices and performances demonstrate alternative social configurations and promote transformative ecopolitical professionalism (see Fig. 5.1). As encapsulated in Fig. 5.1, creation of the festival has not only provided an impetus to sustain cultural heritage through providing new income streams for musicians and their communities, it has also provided a means to address social problems such as gender exclusion, develop educational possibilities, and undertake international research and dialogue. These processes have fostered a collective interest in and commitment to a public pedagogy that promotes transformative change in continuous, iterative, cumulative processes. Riju's festival work has developed "new modes of perceiving and sensing while avoiding to 'teach' an explicit critique," as Schuermans and colleagues (2012, p. 679) wrote on the potential of public pedagogy.

Riju Tuladhar's narrative provides an example of the ways in which a "grassroots" intervention of a musician and music teacher in Nepal can aim at significant social, cultural, and economic transformation through an active agenda for sustainability in a context where caste distinctions, gender inequalities, and opportunity gaps are maintained and sustained through traditional music heritage and festivals. The narrative illustrates the tensions between these traditional structures and music practices,

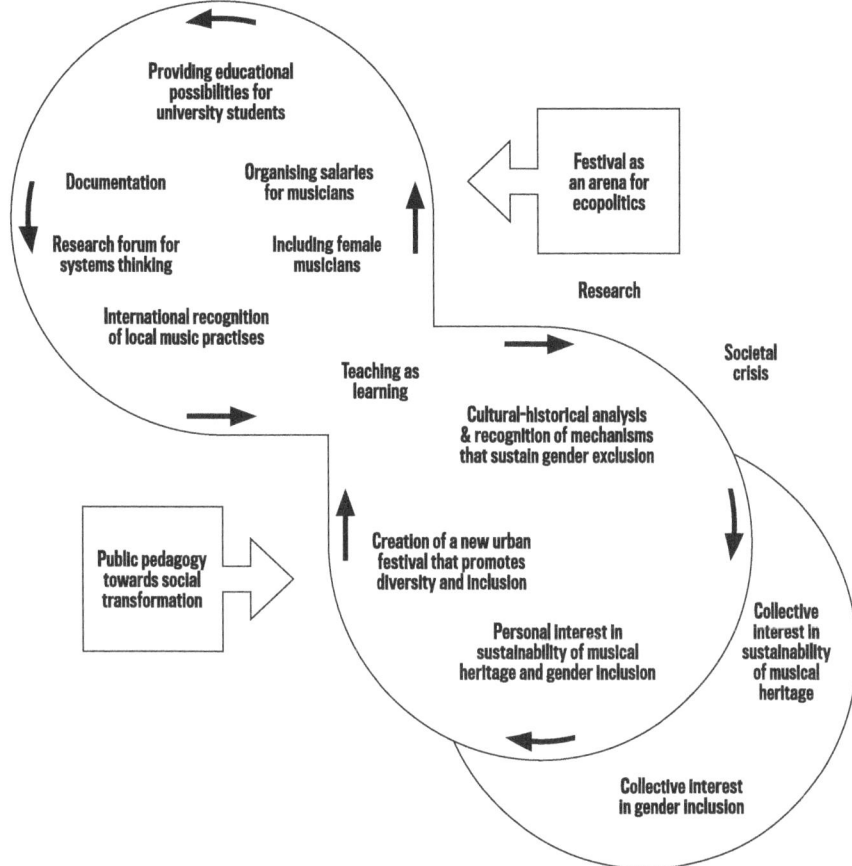

Fig. 5.1 Public pedagogy and transformative ecopolitical professionalism

sustainable practices for music development, and the emerging democratic values, policies, and practices of the Nepali context. In negotiating these tensions, Riju demonstrates the possibilities of alternative modes of thought and actions in acts of public pedagogy that avoid an "explicit critique" and invite community engagement and collaborative action.

References

Bauman, Z. (2017). *Retrotopia*. Polity.
Kansakar, S., & Tuladhar, R. (2020). Enabling grassroots participation in the promotion and preservation of traditional musics: The case of the Echoes in the Valley music festival in Nepal. *Finnish Journal of Music Education, 23*(1–2), 87–93.

Schuermans, N., Loopmans, M., & Vandenabeele, J. (2012). Public space, public art and public pedagogy. *Social and Cultural Geography, 13*(7), 675–682.

Westerlund, H., Karttunen, S., Lehikoinen, K., Laes T., Väkevä, L., & Anttila, E. (2021) Expanding professional responsibility in arts education: Social innovations paving the way for systems reflexivity. *International Journal of Education and the Arts, 22*(8). https://doi.org/10.26209/ijea22n8.

Chapter 6
Boundary Spanning Ecopolitics

Abstract This chapter presents a narrative of Grammy Award-winning musician, Global Ambassador for UNESCO, UNCCD, UNICEF, and the United Nations, and environmentalist Ricky Kej. The chapter explores his pathways towards becoming an environmental ecologist who uses his work as a musician to engage across multiple political and social arenas and agendas. The chapter presents Ricky's work as a form of boundary spanning in which he undertakes public pedagogy to illustrate the connectedness of the world and the need to engage in systems transformation thinking and action to address "wicked problems" such as climate change and its devastating global effects. Ricky's mission to change global practices that negatively impact the world's ecosystems resonates with Zygmunt Bauman's call to "take a moral stance" and act within multiple and diverse spheres of influence.

Re-Envisaging Music Education for Ecopolitical Engagements

Beginnings: Listening, Playing, Questioning

The public biographies describe musician Ricky Kej as a Grammy Award-winning Indian Music Composer and Environmentalist, a UNCCD (United Nations Convention to Combat Desertification) Land Ambassador, a UNESCO—MGIEP (Mahatma Gandhi Institute of Education for Peace and Sustainable Development) Global Ambassador for Kindness, a UNICEF Celebrity Supporter, an Ambassador for the Earth Day Network, and a United Nations Refugees Goodwill Ambassador. Indian Prime Minister Narendra Modi and then French President François Hollande launched his 2015 album *Shanti Samsara—World Music for Environmental Consciousness* at the 2015 United Nations Climate Change Conference. Ricky has been recognised as a Global Humanitarian Artist by the United Nations and honoured by numerous countries. His website notes that "his vast repertoire of work includes 16 studio albums released internationally, over 3500 commercials and 4 feature films, including the natural history documentary 'Wild Karnataka' narrated by Sir David Attenborough" (https://www.rickykej.com/about). His leadership as an environmental activist is recognised globally. And yet, in the music education community, his work as a musician and environmentalist is less well-known.

How does one become an environmentalist? How does one develop ecocultural, ecological, and environmental consciousness? What role does music play in this process? What does environmentalist action in and through music look like? What are the motivations and experiences that prompt taking that pathway through music-making? We met with Ricky in early 2021 to explore some of these questions.

The chronological facts of his life are well documented across a range of websites. Ricky grew up in North Carolina, US, and lived there with his family for the first 8 years of life. He speaks of his father, a doctor by profession, as a music enthusiast who "*had a huge music collection. And not just the regular pop songs of those times like the Bee Gees and Michael Jackson, Elton John, and things like that. But also he had a whole lot of world music, like these African choirs, musicians from Vietnam, musicians from India, of course, Celtic music, and things like that. So, I would listen to all of this music.*" In addition to access to his father's extensive music collection "*there was a guitar lying around at home. There was a really tiny, Sprint's piano at home. And what I would do is ... I just picked it up and I started playing on it. And I started asking people, a lot of questions ... And I do not know why we had these musical instruments at home because my father and mother both would not play any musical instrument.*"

An Early Obsession

Ricky's "*obsession with music growing up*" was not limited to the sounds of music. He describes at a very young age reading "*the credits on the LPs and the 8-track cartridges and cassette tapes*" in his father's collection. "*I would read the credits, and I would play this game; I'd try to find the same names in different albums. And I would try to see who was sharing with which musicians, and which producers, and things like that.*" He describes this "*game*" as important in "*learning about how music is created*," summarizing "*it was not just listening to music but also playing music, and also the creation process itself that was quite fascinating for me.*" While for his classmates the "*visual medium*" of television and cartoons was the focus, for Ricky "*It was always music ... my music system was pretty much the centre of my universe.*"

Ricky's "*passion*" for music continued after the family moved to Bangalore and was evident throughout his schooling to Grade 12. He was a member of a number of school-based bands, mainly as a keyboard player, and represented his school in various competitions. He notes that "*by then, my heart ... was with electronic music, ... more towards the technology in the creation process and the performing process of music.*" Ricky began writing his own music "*in the sixth or seventh grade itself, ... 11 or 12 years old. ... it wasn't writing scores ... or notations. It was just writing lyrics, putting together some chords, and writing my own songs. So, I had already started expressing myself through music by that time."*

Career Decisions: Conflicts and Compromise

"In India, it's in the 12th grade that we have to make a very strong decision as to what we want to do for the rest of our lives. I had made a very strong decision: I wanted to become a musician, and my father wanted me to become a doctor ... there were a lot of ... fights at home because of that ... a lot of drama at home. In India, you cannot NOT listen to your parents. So, at the end of the day, I had to, no matter how strong-headed I was, I had to make a compromise with my father that I would finish off a degree in dental surgery. And [we agreed] once I got that degree then my life is my own, and he would never question me for the rest of my life. So that's the deal I made with him."

Ricky's enrolment at dental college in Bangalore did not put music into the background. If anything, it provided him with the space to pursue a formal education in music. Over the 5 years of dental college *"in the evenings, I studied the London School of Music syllabus (piano) with a tutor ... I did a couple of grades. And I also studied Hindustani classical music with a tutor."* Ricky reflects on the differences between these two courses of study noting that his tutor in Hindustani classical music *"modified my course ... in the sense that he would tell me, 'Okay, this is raag. This is the raag Bhupali.' He would tell me the raag name and then say, 'An example of this is Pink Floyd's,* Wish You Were Here.*' He basically combined both my worlds—very beautiful for me ... it was more of a theory education when it came to Indian classical music."*

Upon graduation from dental college, Ricky gifted his certificate to his father. He notes *"at the end of 5 years, I finished off my degree, and then that is when I became a full-time musician. I did not practise [dentistry] even for a single day."*

Being and Becoming an Environmentalist

Ricky reflects on when and where his love of nature developed: *"my love for music and nature went pretty much hand in hand ... right from the time when I was in America. In North Carolina we lived in a very remote area ... pretty much in the middle of nowhere. And it was in this small city known as Roanoke Rapids, which was on the border of North Carolina with Virginia. We had this huge, wooded area behind our home, and quite a lot of open space. We would have a whole lot of so-called creepy-crawly animals that would come into our home on a regular basis. And my parents would constantly want me to just run away from them. And the knee-jerk reaction was always, from everybody around me, that if you find any of these animals just immediately kill them. Step on them and kill them. I was very fascinated by these animals, and I would be very drawn towards them. And at a very young age, I used to have these existential arguments: I would ask my parents, 'Why do we have to kill them? Why do they exist if they need to be killed as soon as we see them? They must be having some purpose in life.'* ".

Whilst the move from Carolina to India at 8 years of age saw a move from a rural, wooded landscape to the urban cityscapes of Bangalore, Ricky was still drawn to the green spaces of that city, as places of peace. He remarks, "*I used to love hanging around in parks ... that was not real nature because that was manicured gardens and things like that. But nevertheless, for me, I always felt at peace, being in green areas. I believe that most children are that way ... most children are very environmentally conscious, and most children are very ... they are more drawn to love nature and things like that. It's just that we systematically erase that from them as they grow up.*"

These early experiences underpinned Ricky's developing awareness of the interconnected nature of living systems. Reflecting on these he notes, "*And then of course much later on I understood about the ecosystem. And I understood how every single living being is important from the perspective of maintaining the delicate balance of the ecosystem. We need to live in coexistence for our own survival. At a very young age, I used to constantly have these questions. And then, later on, I realised that I'm something called an environmentalist.*"

Becoming a Professional Musician

Ricky traces the beginnings of his professional career as a musician back to his days at dental college at the age of 19 years. He recalls that, "*at that particular point in time, exactly when I was 19, that's when the first FM radio station started in Bangalore. So, I immediately saw an opportunity over there because I realised that these FM channels have ads. They have advertisements playing on there so maybe I can compose those advertisements and I can create some music for them. So, in the evenings, I would be at some sort of a studio, and I would be recording.*"

Ricky's career as a music commercial advertisement composer took off and over the ensuing 13 years he "*created over 3,500 commercials for television and radio.*" He continues, "*there were days when I would do between two to three commercials a day for different parts of the world. I would start off in the morning being briefed on a television commercial which is going to be played in India, then I would finish that off by afternoon. Then in the early evening I would be briefed on something from France, or from Spain, or from Germany. Then late at night it would be New York, or it would be Los Angeles. So, I would barely sleep, and I would constantly keep composing music for these television and radio commercials.*" Ricky composed music for every occasion ranging across "*audio-visual presentation, or television or radio commercials, or an anthem for employees.*"

Shaping a Music Identity

Amidst this busy schedule Ricky also sought to "*make music which I truly believed in on topics that I felt strongly about …. I decided to become an independent musician when it came to my own music.*" In pursuing this topic, he distinguishes between the possibilities offered by the Indian film industry (Bollywood), the commercial music sector, and his own music-making, focusing on the ways in which one's musical identity is allowed free expression or co-opted into the expressions, thoughts, and philosophies of others.

"*India is a dominantly movie industry market, in the sense that in India, 99% of all music content is created by the movie industry. Because our movies have a lot of songs, a lot of movies are musicals. And a movie cannot survive without songs, because the main promotional medium of a movie is basically songs …. That was something I did not want to get into because, I'm doing the commercials anyway … I did not want to get into a whole other industry where I would not be able to express myself. I chose advertising commercials over movies because movies are a credited medium. In movies, you're making music that does not define you as a person. It's basically based on a script, based on the thoughts and philosophies of a director and a script. But it is a credited medium, so your name is attached to it. So, your name is attached to that product, not you as a person. It's not reflecting what you feel. Whereas when it comes to commercials, commercials are not credited mediums, you have no idea who's actually composed music for it.*"

A certain pragmatism also shaped Ricky's decision to work in the "*non-credited*" space of commercials rather than the film industry, a pragmatism shaped by the elements of time, collaborative possibilities, and speed: "*The difference between commercials and television and working on a film is that working on a film takes a very long time, in the sense of composing five or six songs for a movie and doing the score, it's pretty much a year-long process. Whereas with commercials it's just a few hours. In a few hours, you just wrap it up. And you're working with a different creative person, for every particular jingle. So, you're meeting all these new people. For example, yesterday I worked on an Indian folk jingle. Today, I get to work on a rap jingle. Tomorrow, I get to work on a heavy metal jingle. So, you have to be on your toes constantly about different forms of music because you do not know what you're going to be briefed on. And you have to deliver on the brief within, like, half a day or one day. So that's why I had this beautiful network of musicians all over the world of various different genres, various different cultures, various different traditional instruments. If I get a brief, I can immediately make a phone call, and I can tell that person that, 'You guys are booked today because I am going to finish off a composition, and I'm going to send it to you, and you'll have to immediately record your parts and send it back.'* ".

Working in commercial music has not only developed Ricky's capacity to compose quickly, with numerous collaborative partners, across multiple musical genres; it has also taught him to weather criticism. "*Commercials have taught me to have a very thick skin when it comes to criticism, and to take criticism constructively. Because, when you make something, let's say I'm making something for Toyota ... when you're working on a brand like that, and the client is not getting what they want, they do not mince words. And they're not criticising you because they don't like you or because they have something personal against you. Of course, in the beginning, I used to be like, 'What do these people know about music? They know nothing.' But then, later on, you start realising that they have nothing against you. It's just that it's not working for them, because for them a jingle is not beautiful music or not good music. For them, a jingle is correct or wrong. Whatever is working for the brand or not working for the brand."*

Fusion Musical Identity: Breaking Cultural Barriers

Ricky attributes his training in advertising, his exposure to multiple genres of music, and collaborative work with musicians across these genres to his emerging identity as a fusion musician, whose music draws on many musical genres. He describes this training as a workout that has built his capacity to compose quickly and on demand: *"it's almost like a workout, you know? The more you work out at the gym, the better you get at it. And I believe the number of times that I've composed music for advertising gigs—I'd say more than 3,500 times—has made the composition come to me very, very easily, and ideas and melodies come to me very easily."*

Ricky's identity as a fusion musician has led him away from the composition pathways offered in India through the movie industry to seek international collaborations and opportunities. He also sought models in the work of other successful musicians, including Pandit Ravi Shankar. He recalls: *"I went to a Ravi Shankar concert. You know, my first one ... in the Bay Area. And I was very surprised that he was playing pure Indian classical music. And the audience was very representative of the exact percentages of people and cultures that were in America. Like, we were just about 1% or 2% Indian people, and the rest just American people, you know? I was very surprised, and I thought that, 'Okay, this is exactly what I want to do with my life.' That I want to break cultural barriers and I want to take Indian music all over the world."*

The beginnings of Ricky's international recording career were somewhat serendipitous. As he explains it: *"Right at the beginning of my career, like, 1999, 2000, I did my first album. It was called 'Communicative Art,' and it was released with a really small boutique Indian label. A spa store in New York, which was owned by a friend's friend ... found my music to be very meditation and very fusion-ish kind of music. So the friend's friend asked 'Why don't you send me some CDs, and I'll put it in the spa store, even though we don't sell music. I'll just put it there so people who are buying candles, and incense sticks, and yoga mats, and things like that, they might want to buy this music because it may work for yoga. And we'll play it in the store.' I was very excited about it, so I FedExed about 15 CDs.*

And then almost a year later, I got a call from an American phone number. And this person speaks to me and says, 'Hi, I'm Rod. I'm a vice president with Universal

Music, and is this Ricky? Okay, I got your phone number from such and such person. A year back, I had gone to the spa store and I picked up the CD. I've been listening to it constantly in my car for the last year. And I love the music. With great difficulty I found your phone number. Are you signed with anybody?' I said, 'No, I'm not signed with anybody.' So, then he says, 'All right, let's work together'."

From that beginning Ricky developed a parallel career as a fusion musician, eventually making some 12 albums with American labels such as Universal Music, EMI, and Water Music records.

A Moral-Ecological Turn

A Huge Shift: Going Independent

Ricky speaks of a *"huge shift,"* a *"profound revelation"* in his life around 2013. Reflecting on his career in advertising he asked: *"What am I doing with my life? I started thinking about all these commercials that I've worked on. I started realising that these big brands like Google, and Microsoft, and Pepsi, and Coca-Cola, and McDonald's, the brands that I was working with, they have understood how powerful music is to sell their products. They've understood that music is such a powerful medium rather than just a talking ad; so much so that they're ready to spend a few thousand dollars on me to actually compose music for them. They are ready to spend a couple of million dollars to actually air this music on television and radio, because they completely understand that music is luring people not just for communicating a message, but for getting that message deep in the head of a listener. I decided that I needed to explore because there were so many things that I felt strongly about, especially the environment, and social justice. So, I decided that I'm going to stop all forms of commercial music and every single piece of music that I make is going to be about, for lack of a better way to explain, to just make this world a better place in my own small way. So that will be my mission till the day I die. So that's where I am right now."*

That decision coincided with the finalisation of a number of contracts and the realisation of financial independence through continuing ownership of the rights to all his music. He explains: *"in all the commercials that I had done, all the music at the end of the day the rights of all that music did belong to me. I had a publisher in America who was still working those musics. In commercial contracts the music is usually licensed to them for about 1 year, or a maximum of 2 years. And then after that, they have to renew the licence. So, I had a huge library of music and the publisher is constantly repurposing that music for different people. And that publisher still does that. So, I've got a steady income that's coming in through royalties. So that was one thing that kept my mind at ease that at least I'm not going to starve."*

Ricky's 13th album "Shanti Orchestra" was his first independent album followed quickly by his 14th album "Winds of Samsara" which went on to win the Best New Age Album trophy at the 57th Annual Grammy Awards in 2015.

Becoming an Environmental Advocate

The nomination and subsequent awarding of the Grammy had a significant impact on Ricky's life as he explains: "*After I won the Grammy Award, two things happened: my name was called out and I remember very clearly that when I got up the only thing that was coming to my mind while I was walking up to the stage to give my acceptance speech, was that, now I can do music forever, nobody can stop me. I'm going to find a way to do music forever. Because up until then there's always these doubts. But I felt that with this award and this recognition there's no way I can stop doing music now.*

And then our Prime Minister, Narendra Modi, invited me for a private meeting, a 10-minute photo op with the award and congratulating me because I was only the fourth Indian to have won the Grammy Award. What I thought was going to be a 10-minute meeting ended up being an hour-long discussion and it was a beautiful discussion because he realised that I was an environmentalist and that was my passion. And he told me that later on that year he was going to be attending the climate change conference in Paris, that is the COP21. And so he told me that he's going to be having the mainstage for a good 1 hour, and he's going to be launching something known as a Solar Alliance, which he launched over there, which is basically countries with tropical climate coming together as an alliance and producing solar energy. And he said that if you make an album, which includes musicians from all over the world and which is about climate change, and which is about environmental consciousness, he said, I will launch it over there on that stage, and I will make sure every single world leader gets a copy of the album. So, I thought, 'Wow, that's amazing.' And then he just told me while I was leaving his office, he just told me, 'So why don't you just dedicate your life just to this, not just music … become an advocate, help us out, help our government out, help the United Nations, help everyone and just do this for the rest of your life.' And I took that advice very seriously."

Ricky made that album, "Shanti Samsara," with 500 musicians from 40 countries, including musicians from the Royal Philharmonic Orchestra; Philip Lawrence from Bruno Mars; Peter Yarrow from Peter, Paul, and Mary; the Nashville Symphony; and "*a whole lot of musicians from India, basically, every corner of the planet wherever I could find musicians who felt as strongly as I did about the environment we just collaborated with them. And of course, our Prime Minister released the album with the then French President François Hollande. Almost immediately after the release of the album, I ended up performing at the United Nations General Assembly. And that is where a bunch of agencies contacted me for concerts, and also for advocacy, and then I ended up becoming an ambassador with UNICEF, then with UNESCO*

MGIP, and then with UN CCD, that is the United Nations Convention to Combat Desertification."

Being and Becoming an Advocate

Acting on the advice of Prime Minister Narendra Modi, Ricky has used his musical skills and activated his global networks for a range of campaigns. One of the first major campaigns was the WHO campaign BreatheLife. This campaign "mobilizes communities to reduce the impact of air pollution on our health and climate."[1] By linking an environmental problem (air pollution) to its health effects, including bronchitis and lung cancer, the campaign seeks to elevate the importance of national and global action on climate change. As Ricky describes his involvement in the campaign, *"a whole lot of statistics came out through the World Bank that 7,000,000 people die every single year because of air pollution, out of it 600,000 of them are children. … the BreatheLife campaign tried to get countries to sign up on this … even regions like mayors of various counties could sign up and share knowledge and share ideas of how to combat air pollution. The WHO got in touch with me and they said, 'We're going to be starting the UN conference on Monday …. Why don't you come on Sunday, all the world leaders will be here at the headquarters in the Palais des Nations in Geneva, why don't you do a concert, inspire the hell out of these world leaders, so that the next day, they will give us commitments, do something really powerful.' So that's what we did, we put together a whole lot of musicians, we performed some new compositions of mine. At this concert we got all the world leaders, mayors of various areas, and presidents, and prime ministers, and heads of intergovernmental bodies, we got them to stand up, we got them to dance, we got them to sing along with us at this 2 hour concert. The next day, Dr. Tedros invited me for the sessions and I saw that almost every country was giving them commitments. A lot of people actually quoted phrases from the concert itself. As Yewande Awe, Senior Academic Advisor at the World Bank, said at this concert, "Tackling pollution is not the exclusive preserve of academics and experts. There is a place also for artists"*.[2]

Working With and for Children

Ricky works across a range of UN agencies including UNICEF for which he is an MGIEP Kindness Ambassador. He begins his account of this work describing a realisation that *"while I was doing my advocacy for the environment, it dawned on me*

---

[1] BreatheLife Campaign: https://www.ccacoalition.org/en/activity/breathelife-campaign#:~:text= BreatheLife%20is%20a%20global%20campaign,Clean%20Air%20Coalition%20(CCAC).

[2] BreatheLife Concert at the UN Headquarters in Geneva 2018: https://www.youtube.com/watch?v=exQ9p1VZ0lI.

that in India ... the environment is considered to be a thriving problem rather than a survival problem because in India, we've got much bigger problems. Things like climate change even though it's a huge existential problem people do not consider it that way. Because people feel that poverty, education, gender inequality, gender violence, water sanitation, are much bigger problems."

He continues: *"I realised that you cannot take the environment in absolute isolation in India, you need to tackle a whole lot of other problems before you get to the environment. Because if I go to a rural area, which does not have electricity, where people are not productive after 6:30 in the evening because they do not have lights, they do not have water, they don't have sanitation, and no education for their children: if you go to somebody like that and tell them let's make a better world for our children and our children's children but you will have to live in poverty for the next 20 years because you have to figure out this whole solar thing, they're just going to turn around and probably slap you or something, because they're going to be like, 'What about me?' I realised that we need to fix all these immediate problems, these survival problems before we even can think about things that are perceived as being thriving problems. So that's the reason why I decided to work closely with UNICEF, and I'm their ambassador for India."*

Ricky's role at UNICEF encompasses advocacy for children's health and children's rights, and promoting awareness of and action on the 2030 United Nations Sustainable Development Goals (SDGs). He describes the SDGs as a *"beautiful framework for a holistic approach towards solving problems"* which inspired him to create the My Earth Songs project.[3]

The project involved working with musicians Lonnie Park and Dominic D'Cruz to take the complex ideas of the SDGs to compose *"27 extremely simple songs, very easy to learn, with simple chords. And we disseminated them through the education system through textbooks. The publishers took these 27 songs, dividing them from Grade one to Grade 8 based on complexity. And within each grade, based on the thematic element of the songs, dividing them between social studies, general knowledge, and science textbooks. We are in about five million textbooks across India. Each textbook has approximately two to three songs. And next to each song, there's a QR code so parents and teachers can just scan the QR code and the song appears for free onto their phone. And I've put all the songs into public domain so there's no copyright, no restrictions for use and anybody can use them for whatever purposes they want. A couple of schools in Nigeria have adopted them, a couple of schools in South Africa, a couple of schools in Russia, a couple of schools in South Korea, and they're using them in their educational curriculum. Universal Music in the UK released an album version of it. So that's available on platforms like Spotify."*

Ricky describes the feedback received by the My Earth Songs team, including *"schools that come back to us, with stories, that the children have forced them to go zero plastic because you have a song about plastics."* He continues, *"at one particular school in New Delhi where they were constructing a building in an open space, the children protested, 'No, we want the open space, we don't care about the building.'"*

---

[3] My Earth Songs concert: https://www.youtube.com/watch?v=UYBqQ3cNFV8.

The songs have not only motivated children to action in school environments, they have also impacted the home and community. As Ricky explains, "*a whole lot of parents are writing to UNICEF and telling them that our children have become activists at home, and they do not accept chocolates simply because they are in a plastic wrapper ... it's been very nice to see the difference that the songs have made .... I have always believed that the songs you learn during your childhood are songs that you never forget .... If those songs have some amount of morals in them, those morals stick with you forever so that's one of the main reasons why I decided to do this project.*"

Engaging in Public Pedagogy: Affordances and Constraints

Ricky has visited over 200 schools across India to meet students and teachers and further the advocacy of the My Earth Songs project. He explains: "*Whenever I visit a city for a concert, I just call up the publishers of the textbooks and ask, 'Which schools do you think I should visit?' They put together an assembly of all the students and I speak to the students. And it's pretty surprising as to how well-versed the students are about what climate change is and what are the threats of climate change, and what we need to do in terms of climate action and not just through the songs, but in general. By contrast, when I give talks at corporate events and I speak about climate change, they don't seem to know anything about climate change, they just know the term climate change. And then I ask them, 'Do you know what climate change is?' And they're like, 'Yeah, you know, something about the polar ice caps melting,' but they do not know exactly what it is, how it is a threat to us, and how things are compounding. But the children seem to know a lot more.*"

Ricky's work as a public pedagogue is evidenced in the educational thrust of his music-making as he writes songs for varying audiences, giving concerts, and engaging with different constituencies. We asked him what skills and knowledge have helped him to develop that identity and capacity to engage in public pedagogy, and what advice he would give to those doing the same. Ricky commenced by emphasising, first, the need to know both the science of the issue as well as the artistic representations, and, second, of fostering interdisciplinary collaborations between artists and scientists: "*First of all, you have to have a very strong basis in reality and in science because not everything can be just about artistic representations of everything. There has to be the meat of it, in science. I'm very grateful that I met the Dean of the National Institute of Advanced Studies, NIAS, Professor Baldev, who attended two or three of my concerts. At one of the concerts, he just came backstage, and he told me that 'I would love for you to speak to my scientists.' I ended up giving about three or four lectures over the next year or two, to the scientists at the Indian Institute of Science. And then one fine day, he just tells me that, 'I want to make you a professor over here, because what you are doing, essentially, is science communication.'* ".

Ricky's science communication projects at NIAS included a science communication collaboration with scientists on the effects of climate change on the island nation of Kirabati, located in the middle of the Pacific Ocean and consisting of 21 inhabited islands. Ricky begins: "*It's predicted that they're going to be the first country in the world to go completely underwater due to climate change, due to ocean levels rising. I visited the country and spent about 3 weeks there. I spent a lot of time with the president, and with the people, to understand the effects of climate change and who are the victims of climate change. I saw a whole lot of villages which are already underwater and populations that are forced to migrate. I realised that this is a country which is not at fault for climate change, because climate change is not at all of their making. They are a country which is environmentally conscious, they are a country which has no industrialisation, barely any carbon emissions, and they do not take more than they can give back to the environment. It's us developing nations and developed nations which are basically sinking that island, because, at the end of the day, we just have one atmosphere.*

I sat with the professors at NIAS and I tried to understand what are the statistics. Why are they going underwater? What exactly is causing this, how many cubic centimetres of carbon emissions is going into the atmosphere, and from which countries? Then I made the song on Kiribati,[4] *to interpret all of this and to bring an emotional angle. You can throw a whole lot of statistics at people, you can throw a lot of data at people, and scientific data, you can give all your speeches to people, but if you perform a song in front of people or make an audio-visual song, then that goes into the hearts and souls of people. I decided to showcase the children because one thing that the scientists told me is about the number of years that Kiribati has. Based on various calculations around the 1.5 degrees centigrade target that the Paris Climate Change agreement has mandated to other countries Kiribati has maybe 30 or 40 years. The immediate emotional thought that came to my mind was that the children won't be able to grow up in that country. The children are going to have to migrate, they won't be able to grow old in their own country. That is what the song is about, we showcase a whole lot of children in the song and through the lyrics, we talk about how these people will have to leave their country, they won't be able to grow old in their country, and their culture is going to go extinct. Basically, I am taking ideas, scientific ideas, and interpreting them in a more emotional sense.*"

In elaborating on his public pedagogy, Ricky describes his concerts as being of two kinds: the top-down approach and the ground-up approach. "*The top-down approach is when I perform to world leaders, and when I perform at the UN headquarters, or to Fortune 500 companies, or Fortune 200 companies, where it's smaller, intimate audiences of, let's say, about 200 or 1000 people. There the messaging is more about "You guys can make a huge difference" and, 'It's your responsibility.'* ".

The other kind of concerts that I do are basically the ones at music festivals where my largest audience ever has been 88,000 people. I regularly perform to audiences of 10,000 and 20,000 people where the messaging is completely different.

---

[4] Song for Kirabati: https://www.youtube.com/watch?v=LGwpcH79bps. Further information at: https://www.youtube.com/watch?v=NToedyJPJjY.

There I talk about these problems that we are facing, especially the environmental problems. The biggest threat to us is basically the constant thought that somebody else will make a difference, that governments will make a difference, intergovernmental bodies and leaders and politicians when the truth is that the only way we can bring about difference is basically through behavioural change, that we need to empower ourselves." At these concerts, Ricky seeks *"to empower every single person to believe that they have the power to bring about change by just bringing about small incremental changes within their own lives."*

Ricky describes his dual strategy of top-down and ground-up approaches as one that holds leaders accountable in an effort to move them from being followers of popular opinion to *"taking strong decisions and long-term solutions towards their constituents."* He emphasises the role of artists in a ground-up approach, *"not just musicians but through filmmakers, through sculptors, through dancers, because that's the only way that you can change public consciousness, because we need to elect leaders who will make strong decisions."*

Ricky stresses that focusing on changing public consciousness does not mean relinquishing creative ideals. He continues, *"I try to make my concerts as entertaining as possible, because that's the most important thing, that the music has to be entertaining. That's the only thing that draws people and the music has to be really good and it has to be entertaining; it has to be powerful."* However, he acknowledges the constraints of taking a proactive political stance describing how *"a lot of artists don't want to collaborate with you ... because a lot of artists feel that if they become ... if they are perceived as being very issue-based artists or they're perceived as being after a certain cause that they lose a whole lot of their audience. I've had a lot of artists who have refused to collaborate with me because of this. Because their managers or the agents tell them that. Sometimes the artist is very much into it. They feel like, 'I totally believe in this, in elephant conservation. I love elephants, let's do this together.' But then 2 or 3 days later I get an email from the agent saying 'No, they're very busy or whatever.' I realise that the agents do not want them to do this because apparently they believe, and probably it's true too, that a whole lot of their fans apparently are completely alienated as soon as you get to any kind of causes. And especially if you look at climate change. Climate change is considered to be a science everywhere in the world except for America. In America when I speak about climate change on stage it is considered to be a political term, or it's considered to be something you would probably believe in or not believe in, when it is a fact everywhere else in the world. So, if I speak about climate change on a stage in America, I'm immediately considered to be a Democrat."*

These thoughts lead Ricky to his third recommendation for public pedagogy: Stay apolitical. He comments *"I do have my political preferences personally, but you will never see that on my social media, or you will never see that at a concert, you will never see me endorsing a particular candidate because I believe that whoever has been democratically elected, I will work with that person. And that's why I love the United Nations because that's exactly how the United Nations functions. You'll never see any of their employees protesting outside a government office because they just believe in whoever the government is, however horrible they are, or however good they are, they are just going to work with them. And that's how I base my thoughts, that no matter how much I dislike a particular politician, I will figure out what are their motivations and I'll work towards that motivation so that they do the right thing."*

Everything is Connected: Systems Transformation for Environmentally Sustainable Futures

Ricky's song-making and public pedagogy foregrounds building understanding of human impact and the environment and the interactions between the complex ecological, social, economic, and cultural systems in which we live. To illustrate this point, Ricky describes another science communication project he has developed with a professor at NIAS working in elephant conservation and understanding elephant behaviour. Ricky was struck by the conundrum of elephants being regarded as gods in India, such as the elephant god Ganesha, yet at the same time being treated poorly.

"I created one song which asks, 'Why is this hypocrisy?' Every time you start something new on an auspicious occasion you want to pray to the elephant but at the same time, you're treating the elephant so badly. But the second thing, which is more important, was cause and effect. A year back in India there was this huge incident in the state of Kerala where an elephant had consumed this fruit which had some explosives in it. I'll give you a little bit of background. There's a huge thing called human and elephant conflict in India. We still have these large mammals in India, and we have urban forests where people are still encroaching forests. The elephants end up going into agricultural spaces to find food because their own habitats are degraded. And when they get into these agricultural lands, the farmers, to protect their livelihoods, take really terrible options. They used to do these electric fences, and then electric fences were banned in India, because elephants would die on these electric fences. And then what they started doing is that, illegally, they started having these fruits like pineapples, which the elephants love, and they would put explosives in it and try to kill the elephants, which is absolutely horrible. But they used to do that to protect their livelihoods and protect their families.

This one incident happened where the elephant ate the pineapple fruit and she was a pregnant elephant. Her entire mouth exploded and she was dying, but she decided not to harm anybody and she walked right up to a river and she stood in a river and she died standing there. It created a huge rage in India and all over the world where everybody said those farmers need to be punished and they need to be put under a firing squad and they need to be hung. But then I created a song called 'Love divine'[5] where what I did was that I spoke about all you city people you need to look inwards about this problem because, at the end of the day, when you do not have the latest iPhone because it goes out of stock, you're like why aren't they building enough iPhones? Or the latest gadget when it comes out, you're like why aren't they

---

[5] Love Divine video: https://www.youtube.com/watch?v=7eUxV9K4S-4.

creating enough of it? Or if you do not have electricity for about 5 minutes you're so inconvenienced, but where is all the coal coming from for that electricity? It's coming from mining the forest. And where is all that manganese and zinc and iron ore, and lead and silica coming from for those iPhones? They're coming from mining the forest and when you mine the forest the elephants have nowhere to go because their habitats are destroyed. So, they walk out of the forest and they go to the people on the frontlines of the forests, that is the villagers, and they start destroying their property. They sometimes cause damage to human life and sometimes the children are killed in the process.

So, they have to protect their families, they have to protect their agricultural lands because of the problems that you have created by mining these forests. Of course, these people need to be punished for putting the explosives into the fruits because that's really bad. But at the same time, you have to look inwards that the cause of this problem is basically you. The song starts, 'Mirror, mirror on the wall.' It's all about 'look at yourself,' and, 'the pain and suffering that you inflict on nature is pain and suffering that will eventually be inflicted on you.' That's the idea behind the song. That is an example of how we need to interpret these situations of cause and effect, and showcase to people that the problem is much bigger than just a knee-jerk reaction."

Future Challenges

We asked Ricky to consider the implications of his work for the future and what lessons might be drawn for music education. He is a strong advocate of music education for all. "*I think it's very important to have a music education ... not everybody needs to become a professional musician, but I think it's very important for people to appreciate good music. I believe that that's the only way that the art-form itself can be elevated.*" Thinking beyond music education Ricky suggests that the major problem for sustainability is overconsumption. "*As Mahatma Gandhi, who was pretty much my role model, said 'The Earth can provide for everybody's need but not everybody's greed.' That's a very strong phrase and that's something that I pretty much lived by.*" He also references his commitment to vegetarianism and using public transport. He continues "*It's very important to practise what you preach because people see through that very easily. For a musician, it's very important that they do not do advocacy based on what a friend has told them, that this is cool. You know, everybody's talking about the environment, let's make a song about that or everybody's talking about Black Lives Matter, it's trending on social media so let's make a song about that, or whatever. It has to be something that you truly believe in, something that, you know, you yourself practise, that's very important. And I believe that the music will come automatically, that's what happens with me.*"

Boundary Spanning in Ecopolitical Professionalism

Ricky Kej's work as a public pedagogue began with a commitment to using his music-making to advocate for greater awareness and action on climate change. He acknowledges, "*I've widened my scope beyond the environment*" as the realisation of the interactions between the complex systems in which we live requires action on multiple levels including a commitment to effecting systems transformation. He provides strong lessons for those wanting to engage in public pedagogy: know the science, foster interdisciplinary collaborations between artists and scientists, be involved in science communication, and stay apolitical in terms of party politics. His use of both top-down and bottom-up approaches to engaging with audiences provides a model for a transformative ecopolitical approach to music education and engagement, one which empowers everyone "*to believe that they have the power to bring about change.*"

In a systems thinking frame, Ricky's work can be described as a form of boundary spanning (see Fig. 6.1). In a world where systems boundaries are becoming more explicit due to increasing professional specialisation, individuals and groups now need to search for ways to connect and mobilise themselves across social and cultural practices to avoid fragmentation (Hermans & Hermans-Konopka, 2010). Then the challenge in education and professional work is to create possibilities for participation and collaboration across a diversity of sites, both within and across institutions (Akkerman & Bakker, 2011). Boundary spanning is therefore necessary in changing institutionalised and societal situations and for systems transformation to take place; however, it is also often resisted (Ilmola-Sheppard et al., 2021) because such work challenges the established boundaries of professional responsibility. In music institutions, boundary spanning can destabilise the shared understandings of the purposes of the institution and be experienced as a risk, or a threat (Väkevä et al., 2022). However, recognising these risks and threats might be seen in a positive light as the beginnings of exercising our moral imagination, of envisioning the possibilities and opportunities of systems transformation, and having the courage to embark on change processes and enter the unknown.

Ricky has worked across a diversity of sites and divides, including those of career choices (dentist or musician), musical work choices (non-credited commercials rather than credited identification with a film industry profile), musical genres, individual versus collaborative projects, and transitions that cross boundaries of intersecting professional systems. He works across boundaries between professional music-making and global policy organisations, as well as education, in order to facilitate communication and multi-directional knowledge transfer. In this work his performance spans audiences from schools, to stadium concert arenas, to global political arenas. Ricky acknowledges that his is a unique position, one afforded through the benefits he has accrued from a successful commercial career that has enabled him to live his values and confront wicked problems. Ricky's mission to change global practices that negatively impact the world's ecosystems might be viewed through the lens of Bauman's (1995) call to responsibility: "To take a moral stance means … that

Fig. 6.1 Boundary spanning between local and global level systems

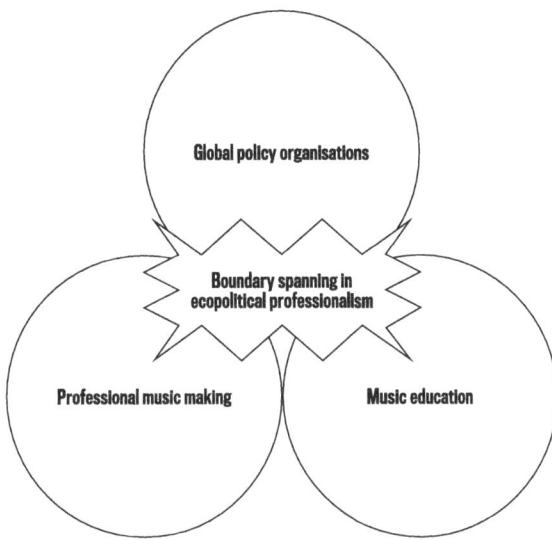

if I have not done it it might not be done at all, and that even if others do or can do it this does not cancel my responsibility for doing it myself" (pp. 267–268).

References

Akkerman, S. F., & Bakker, A. (2011). Boundary crossing and boundary objects. *Review of Educational Research, 81*, 132–169.

Bauman, Z. (1995). *Life in fragments: Essays in postmodern morality*. Wiley-Blackwell.

Hermans, H., & Hermans-Konopka, A. (2010). *Dialogical self: Positioning and counter-positioning in a globalizing society*. Cambridge University Press.

Ilmola-Sheppard, L., Rautiainen, P., Westerlund, H., Lehikoinen, K., Karttunen, S., Juntunen, M.-L. & Anttila, E. (2021). *ArtsEqual: Equality as the future path for the arts and arts education services*. https://urn.fi/URN:ISBN:978-952-353-043-0.

Väkevä, L., Westerlund, H., & Ilmola-Sheppard, L. (2022). Hidden elitism: The meritocratic discourse of free choice in Finnish music education system. *Music Education Research, 24*(4), 417–429. https://doi.org/10.1080/14613808.2022.2074384.

Chapter 7
Undertaking Systems Transformation Through Ecopolitical Professionalism and Public Pedagogy

Abstract This chapter summarises the key arguments of the book and discusses the implications for music and music education institutions and organisations including schools and higher music education and the professionals who learn and work in these settings. The chapter returns to the need to move beyond path-dependency of organisational contexts in music education in order to undertake systems transformation, develop ecopolitical professionalism in music education, and engage in public pedagogy. Through a discussion of the ecopolitical professionalism of each of the narratives (Tuulikki Laes, Riju Tuladhar, and Ricky Kej), the chapter considers the lessons in public pedagogy offered by these three and emphasises the need for moral imagination to engage in boundary spanning systems transformation.

This volume grew out of a concern for the relevance of contemporary music education when considering the global challenges we face in the twenty-first century: what have been defined as "wicked problems," the seemingly intractable dilemmas that are increasingly shaping and defining our worlds. As music educators, we have observed the ways in which the field has considered its relationship to these wicked problems and its capacity to respond. The global COVID-19 pandemic has been, is, and will continue to be a world-changing experience. One of the key lessons that it has provided is that no profession, not even music education, can claim to be immune to larger societal changes. There is an increasing need for the field to analyse its doings and not-doings beyond musical outcomes in relation to society and the public sphere.

For us, it has become apparent that our responsibility as educators precludes us from the easy solutions of turning away, of declaring such problems are someone-else's responsibility and concern. At the same time, we have observed how the field of music education is increasingly worried about its own position in turbulent neoliberal, economically driven societies (Woodford, 2019) in which more-of-the-same advocacy narratives (more time, more training, more money to maintain existing practice) and technical rationality frame all questions and problems as ones only of musical quality and sustainability (understood here as conservation). We have proposed an ecopolitical and moral shift in the ways in which we conceptualise, envision, frame,

© The Author(s), under exclusive license to Springer Nature Switzerland AG 2023
M. S. Barrett and H. M. Westerlund, *Music Education, Ecopolitical Professionalism, and Public Pedagogy*, SpringerBriefs in Education,
https://doi.org/10.1007/978-3-031-45893-4_7

and enact music education as a profession that engages with the difficult dilemmas we face in contemporary societies. Such a transformative ecopolitical shift recognises that music education operates not just as cultural but also as social systems. These systems can exclude as well as include, tacitly maintain inequalities, as well as actively promote equity and social justice and, colonise through imperialist professionalism, rather than engage with ruptures and boundary crossing. Ecopolitical professionalism in music education asks for courage in identifying the potential for good or ill inherent in these systems and ethical and moral professional responsibility that engages with critical systems thinking and systems reflexivity. Such courage and ethical and moral professional responsibility prompts the profession to "see the whole picture" (Senge et al., 2008, p. 23).

In summary, ecopolitical professionalism in music education is a shift from: a narrow *ego*-logical and technicist rationale for music education to a more holistic and *eco*logical systems thinking; and, an inwardly focused self-sustaining rationale to one which recognises the ethical and moral responsibility in the intersecting and connected ecosystems (Folke & Berkes, 1998)—including the natural, social, political—in which we live and work. To engage consciously in ecological thinking and acting is to interrogate "the ways in which modernist Western epistemologies of individualism and mastery legitimate the subjugation and exploitation of peoples" in order to better understand how "place and location" shape, celebrate, and support equity and diversity (Barrett, 2012, p. 207). This conscious engagement is a form of ecopolitical action that opens a plethora of challenges, yet, as we have argued, also offers new transformative possibilities for music education.

Ecopolitical Systems Thinking in Action

The three narrative accounts presented in this book illustrate ecopolitical systems thinking in action in very different ways. Despite these differences, each demonstrates an ethical and moral responsibility that recognises the interconnectedness of individuals and communities, and the cultural, social, political, educational, natural, and material worlds they share. A commitment to ecopolitical work thus involves working "across scales" (Maris, 2015) and recognising the relationality and interconnectedness of all things in the world (Morton, 2018). Importantly, such work involves moving beyond romanticised notions of sustainability as conservation and preservation of past practices to a critical examination of these practices in order to reveal and address inherent inequities, exclusions, and discriminatory practices. Ecopolitical systems thinking in music education exposes and addresses those unintended and unwished consequences of music and music education practices. The narratives further illustrate how the move from an ego-logical to an eco-logical focus supports music educators to develop an ecopolitical public pedagogy as a central component of their practice.

Tuulikki's ecopolitical systems thinking (see Chap. 4) focuses on the opportunity gaps in institutional music education in Finland in order to create new intergenerational possibilities that reposition those who have been diminished by their circumstances or pre-conceived notions of their capacities to engage in music learning and music-making. She engages in a form of non-judgemental feedback and action that recognises how linear and ageist understandings of music learning in formal contexts prevent older adults from entering music education, or families from learning and making music together. Tuulikki's reflexive systems thinking and action prompts consideration of multiple perspectives and the lifewide and lifelong consequences of what we do as music educators.

Riju's ecopolitical systems thinking (see Chap. 5) focuses on transforming the perceptions and capacities of communities in Nepal to engage in whole-of-society shared music-making instead of ethnically defined practices that exclude women and girls. The Echoes in the Valley festival provides spaces for community interaction and intentionally changes the relationships by including females in public performance and challenging the discriminatory values of traditional musical practices. Underpinning Riju's and his colleagues' work is a strong systems awareness of how they, as the country's first generation of professional musicians, need to work to sustain the country's vulnerable musical heritage while at the same time recognising that its historical, patriarchal, and caste-related social structures must change. They bear a responsibility to maintain and sustain the country's numerous forms of cultural heritage while developing new approaches that disrupt the discriminatory values and practices of this heritage and provide opportunities for all constituencies to access and participate in music learning and engagement.

Ricky's ecological systems thinking (see Chap. 6) occurs at "glocal" levels where he uses his substantial social, cultural, and economic capital to bring together musicians in joint ventures that speak across levels of society from individual relationships, through to national and international conversations focusing on addressing global wicked problems. His work as a "boundary spanner," illustrates the potentialities of engaging with the ramifications of international agendas for transformation such as the United Nations' Sustainable Development Goals (2014). He works on this agenda at local levels, for example through the songs and texts developed to build school children's knowledge and capacity for action on wicked problems, through to global levels where he engages with political leaders and policy makers. His understanding of a systems approach, where "everything is connected" works to strengthen the interconnected nature of the 17 Sustainable Development Goals, rather than paying selective attention to what is least inconvenient to implement.

Ecopolitical professionalism in music education thus challenges an "imperialist professionalism" that assumes all music educators have "the onus to conform to the academic practices privileged in Western institutions" (Tran & Nguyen, 2015, p. 962). Rethinking professionalism in music education moves the profession beyond technical rationality, standards, strictly imposed codification of rules of practice, methods, and models, to reinstate the moral and ethical imagination and responsibility of the music education professional. Each of the narrative accounts are intended to encourage the field to consider the possibilities of re-narrating our professionalism against our situational and contextual moral ecologies; to hike horizontally and cross boundaries that often restrict transformational action. The idea is, of course, not new; we are reminded that Small's aspiration for "musicking" was that it generate new social realities. Such new social realities are only possible if we are able to rethink the social relationships in musical practices, not just focus on the musical outcome and recreate existing realities. In this sense, musicking also refers to a fundamentally future-oriented praxis in its critical mode (Westerlund & Partti, 2018), which recognises that our current practices may no longer be ethically and politically justifiable.

The three narrative examples further illustrate the importance of professional autonomy within a system; a kind of moral and ethical stance in relation with our worlds "without occupying the centre of the world" (Biesta, 2017, p. 9). Tuulikki's, Riju's and Ricky's professional stances provide examples of complementary approaches in which working from the grass roots can change attitudes and values, and transform policy, practice and politics in local as well as global contexts. The narrative accounts illustrate how transformative ecopolitical systems thinking can intertwine and reconfigure music and music education practices and in so doing not only transform music and music education professionalism but also contribute to social and societal change.

Our intention in presenting these narratives is to demonstrate the diverse possibilities when engaged in ecopolitical systems thinking; these narratives are *not* meant to be devices for constructing models for music education professionals. Rather, they are intended to encourage and inspire music educators to cultivate their moral imagination, to develop and activate agency in order to engage in transformative ecopolitical practice, within complex contemporary societies. These narratives illustrate how agency and taking responsibility in what matters requires the capacity to navigate between multiple discourses (Barnett, 2015) and negotiate in the "jungle of plural expectations and commitments" of professional life (Solbrekke & Sugrue, 2011, p. 22; see also Westerlund et al., 2021a, 2021b). The narratives illustrate how musicians and music educators, as with other professionals, need to integrate multiple values in their work and renegotiate their professional boundaries when "hiking horizontally" and encountering and engaging with divergent rationalities, epistemes, and moral understandings.

Moving Beyond Path-Dependency for Transformative Music Education Organisations

Hiking the horizontal in social ecologies with an ecopolitical mindset assumes "the nonlinearity, unpredictability and recursivity" of the dynamics of politics, educational practice, and methodology (Biesta & Osberg, 2010, p. 2). However, much of music education takes place in institutional settings where the guiding ideal is to create linear progress, conformity, and continuity. This ideal tends to shape higher music education and music teacher education as secure and stable educational systems that must be preserved and sustained rather than ones which might respond experimentally and imaginatively to changing societal needs and circumstances. This ideal has been claimed to lead to an organisational silo effect (Tett, 2015); path-dependency, understood as organisational stagnation with established norms and principles (Folke, 2006); and a lack of resilience or "ability to cope with and adapt to external pressures" (Sjöstedt, 2015, p. 22). Recognition of the wicked problems the world faces has revealed how customary norms and principles are unable to deal with them; rather, we need systems understanding and critical systems heuristics (Ulrich & Reynolds, 2020) to initiate systems transformation.

Artists and their working practices are often valorised as models for organisational innovation and imagination. For instance the leadership and strategy researcher Sandra Waddock (2021) writes:

> If leadership is about translating vision into reality, and if artists can envision or 'see' what others find hard to or do not yet see, then artists can help us all 'see' into the future that is needed now, particularly where system transformation is needed. Today's many crises—including climate change, racial and ethnic injustice, biodiversity loss, and growing inequality, not to mention the still raging Covid 19 pandemic as this piece is being written—spotlight the cracks in many economic, business, social, and political systems… System transformation is needed on many fronts, particularly around economics and new economies that do not rely on outdated and unwise assumptions about self-interest, continual growth, and dominance of markets and financial wealth over human and natural interests.That is where both leadership and art come in. (n.p.).

Such a view of artists as future-oriented and visionary models of leadership is promising. Nevertheless, this view presents a romanticised view of the art world and its practices, in which the arts systems that shape this world are taken-for-granted as benign and incapable of negative outcomes. The music field in particular, can be resistant to transformational systems change that demands new responsibilities beyond the technical rationale (Ilmola et al., 2021) and a critically reflexive analysis of the systems that shape the field.

Organisational researcher Peter Senge (1990) argues that organisations must be seen as "learning organisations" in order for them to be able to effectively transform themselves. He recommends five disciplines that system leaders should follow in order for such continuous self-transformation. These are: systems thinking, personal mastery, mental models, shared vision, and team learning. Senge identifies seven constraints that can affect leaders when leading a learning organisation and that may be familiar particularly in long-established music education contexts and higher music education. These constraints are: the "I-am-my-position" syndrome, the "The-enemy-is-out-there" syndrome, the illusion of taking charge, a fixation on events, the parable of the boiled frog, the delusion of learning from experience, and the myth of the management team. The I-am-my-position syndrome occurs in music education when professionals in organisations externalise any transformative ideas by taking a "not-my-job" attitude or when institutional leaders stop transformative ideas from being actualised by asserting that these objectives are not part of the job of music educators. In the the-enemy-is-out-there syndrome, professional music educators blame other people and protect themselves when faced with negative consequences or unexpected challenges. This kind of organisational culture is reactive instead of proactive. The illusion of taking charge refers to undertaking "quick fixes" to problems that have no lasting effects and focusing on single events as a remedy to wider problems. By the parable of the boiled frog, Senge warns against not being aware of the external forces in society: A frog does not notice when the temperature slowly increases, until it gets boiled. Senge warns about leaning too much on our experience as we cannot necessarily lean on our past cognitive structures and mental models. Finally, the myth of the management team in this list of constraints refers to the common convention of relying on senior management to solve all problems.

As a remedy for these organisational constraints, Senge recommends developing the idea of shared learning, shared visions for organisational improvement, and collective leadership. For music educators, this is manifested in a willingness and capacity to identify and address root causes—in other words, develop systems thinking and systems reflexivity (Westerlund et al., 2021a, 2021b). Instead of using linear thinking with a growth mindset, institutional transformation requires hiking the horizontal towards context-specific solutions, thinking through long-term change, and learning from each and every event step-by-step.

Lessons in Public Pedagogy for Ecopolitical Professionalism in Music Education

An ecopolitical understanding of professionalism in music education widens the perspective of an educator beyond the genre-related musical principles and skills transmission that currently forms the centre of professional education in music and music education. It challenges a narrow technical rationality as *the* focus of professional reflexivity and suggests attending also to wider horizontal interconnections that tie music education as a phenomenon to emerging ecopolitical concerns. An ecopolitical understanding does not reduce music education to instances of individual learning and performance to simply manifest one's knowing-in-action. Rather, the suggested ecopolitical professionalism expands professional responsibility to embrace the social relationships and situated settings in which it occurs, in order to recognise how music education contributes to "social and cultural geographies of place and identity" (Schuermans et al., 2012, p. 5). Such a recognition involves the acknowledgement that all action in the public sphere is a form of pedagogy, whether this is consciously undertaken or not. As we have argued earlier, taking an ecopolitical stance, with its commitment to addressing larger societal issues, embraces public pedagogy as a conscious form of action. We suggest that this action aims to reconfigure the practice of music education and articulate—theoretically and practically—a repositioned music education for complex societies by exemplifying inclusionary, anti-racist, and decolonial social reconfigurations (to name a few) that contribute to the transformation of attitudes and relationships in society.

Christopher Small (1999) asked, "What does it mean when this performance takes place at this location, at this time, with these people taking part?" (p. 13). This question prompts further ecological questions concerning who is not participating and why not, how can music education create new spaces and relationships, and what "place" means for those involved in music-making. Critical and expanding ecological thinking can, for instance, lead to reviewing and renewing school rituals as social configurations. We may then acknowledge how school rituals "carry cultural codes that inscribe both the 'surface structure' and 'deep grammar' of school culture" and "shape students' understanding of the world and themselves" (Nikkanen & Westerlund, 2017, p. 117). In doing so, we bring to the surface the ways school rituals and generational performance of these rituals shape the moral ecology of schools and function as forms of public pedagogy. School, as a miniature of society, as Dewey (1938) pointed out, engages students not just in learning discipline-specific skills and knowledge, but also leads to collateral learning which is "derived from the social milieu of educational life" (Schubert, 2010, p. 10). This learning through public pedagogy may be conscious or unconscious, reinforcing existing beliefs, values, and practices; a concept encapsulated in the concept of the "hidden curriculum" (Jackson, 1968). Through engaging in transformative ecopolitical systems thinking the notion of a "hidden curriculum" is expanded and made explicit. Public performance in music for example, might then be viewed as more than a presentation of individual capabilities and knowing-in-action, and become transformative social action that seeks to challenge existing social hierarchies, and collectively "perform social difference" (Laes & Westerlund, 2018).

As Riju's work in the context of Nepal illustrates, teachers and musicians can resist reproducing unwanted social hierarchies and exclusions found in society and bring alternative possibilities and structure to the public sphere. The school creates its own public ecopolitical frameworks that engage with multiple epistemic moral frames, values, and policies. Within these structures, an ecopolitical music education can, as scholars of public pedagogy suggest, consider the pedagogical potential of alternative sites such as "neighborhood projects, art collectives, and town meetings" and other public spaces "that provide a site for compassion, outrage, humor, and action" (Brady, 2006, p. 58). Among the lessons derived from the intergenerational public music-making that Tuulikki's nonprofit company organises in urban neighbourhoods is that music educators do not necessarily need to be limited to the conventional places and spaces that are given to them; it is possible to create new (public) places that support and promote public pedagogy to transform public opinion and institutional beliefs about for whom music-making is made available as well as what music education is for.

As a public figure, Ricky understands his professional ecology not simply through music-making, but also as crafting "public narratives" (Ganz, 2011). Public narratives are understood here as *a craft of framing* key messages, through which professionals enact their moral resources, and a process of accessing, articulating, and communicating shared values with emotional content (Ganz et al., 2022). By conceptualising values as affective commitments, Ganz and colleagues (2022) proposed that "public narrative is constructed through the articulation of specific narrative moments (stories of hurt (why I care) and hope (why I can act) that communicate one's moral resources)" (n.p.). This kind of framing of values is often implicit in musical performance. However, as Ricky's narrative shows, public narratives in music can also be consciously constructed to highlight ecological values and a stance in a world where it has become increasingly clear that all professionals should care and act responsibly. The implication here is that preparation to engage in ecopolitical professionalism in music education should include education in the ways in which we craft public narratives for our work.

Viewing music as a public phenomenon, therefore, asks the teacher to see the whole picture as Senge and colleagues have suggested (2008). This, as we have argued, can lead to understanding music and music education in the public sphere as public pedagogy, "as a potential site for social justice, cultural critique, and reimagined possibilities for democratic living" (Sandlin, 2009, p. 3). Such work requires preparation, for example, as Robert Baron (2021) suggests, through incorporating "public practice as a core component of all academic training" (p. 554) in higher music education.

Concluding Remarks

Change is inevitable and unavoidable. It can be incremental or happen swiftly, sweeping us all through unexpected and unplanned events that recast our worlds in previously unimaginable ways as the global events from the 2020 pandemic onwards have so powerfully illustrated. Such change moves us away from teleological views of progress that focus on achieving planned and desired goals and outcomes, to recognise the unpredictability, fragility, and vulnerability of our world and the need for new systems of thinking and acting. Schön (1983/2016) reminded us that in naming "the things to which we will attend [we] *frame* the context in which we will attend to them" (1983/2016, p. 40). When we *name* and *frame* a purpose of music and music education as transformative ecopolitical professionalism enacted through systems thinking and public pedagogy, we open up the potential for music education to address the wicked problems of the world and recontextualise its work as locally and globally responsive and responsible. This renaming and reframing demands that we ask entirely new questions of ourselves as professionals. Such work requires courage to engage with processes of systems transformation and undertake boundary spanning across multiple diverse ecologies. As this book has aimed to illustrate, such work also offers new possibilities and opportunities for professional work in music.

This book is intended as a starting point for our profession to engage in a critical discussion concerning the purpose and rationale of music education in an increasingly complex world. We have drawn on the voices of three musicians and music educators whose work exemplifies ecopolitical professionalism enacted through systems thinking and public pedagogy. Their voices illustrate the shift from the Ego-Logical to the Eco-Logical in music and music education professions, as we have presented in this book. Our approach has been interdisciplinary in nature, drawing on disciplines beyond music and music education including those of systems theory, sociology of professionalism, ecological thinking, public pedagogy and educational philosophy. It is to this last that we return, in particular to the philosophy of Maxine Greene. Greene (2008) prompted us to consider "things as if they could be otherwise" (p. 18) and exhorted us to "open the windows of consciousness to what might be, to what ought to be" (p. 18). For Greene, the liberating effects of imagination "allows for empathy, for a tuning in to another's feelings, for new beginnings in transactions with the world" (p. 18). As did Greene, we emphasise the need to develop our social and moral imagination, a capacity for wide-awakeness, and empathy. Committing to a transformative ecopolitical professionalism and enacting public pedagogy in music education provides us with opportunities not only to consider "things as if they could be otherwise". Such a commitment can move us to action that is ecologically aware, responsible, and transformative; of ourselves, of our field, and of the worlds in which we live.

> This made me think.
> I am just a drummer. I've always thought that it would be enough.
> But if someone asked me to join in a project that had some specific social or political purpose, I would be happy to join in. Yes, I could do it.
> But I've never thought about this before.

References

Barnett, R. (2015). The time of reason and the ecological university. In P. Gibbs, O. H. Ylijoki, C. Guzman-Valenzuela, & R. Barnett (Eds.), *Universities in the flux of time* (pp. 121–134). Routledge.
Baron, R. (2021). Theorizing practice and practicing practice–Public folklore in US higher education. *Slovenský Národopis, 69*(4), 552–569.
Barrett, M. S. (2012). Troubling the creative imaginary: Some possibilities of ecological thinking for music and learning. In D. J. Hargreaves, D. Miell, & R. A. R. MacDonald (Eds.), *Musical imaginations: Multidisciplinary perspectives on creativity, performance, and perception* (pp. 206–219). Oxford University Press.
Biesta, G. (2017). *Letting art teach.* ArtEZPress.
Biesta, G., & Osberg, D. (2010). Complexity, education and politics from the inside-out and the outside-in. In D. Osberg, & G. Biesta (Eds.), *Complexity theory and the politics of education* (pp. 1–3). Sense Publishers. https://doi.org/10.1163/9789460912405_002.
Brady, J. F. (2006). Public pedagogy and educational leadership: Politically engaged scholarly communities and possibilities for critical engagement. *Journal of Curriculum & Pedagogy, 3*(1), 57–60.
Dewey, J. (1938). *Experience and education.* Macmillan.
Folke, C. (2006). Resilience: The emergence of a perspective for social–ecological systems analyses. *Global Environmental Change, 16*(3), 253–267.
Folke, C., & Berkes, F. (1998). *Understanding dynamics of ecosystem-institution linkages for building resilience.* Beijer Discussion Paper No. 112. Beijer Institute.
Ganz, M. (2011). Public narrative, collective action, and power. In S. Odugbeni & T. Lee (Eds.), *Accountability through public opinion: From inertia to public action* (pp. 273–289). The World Bank.

Ganz, M., Lee Cunningham, J., Ben, E. I., & Segura, A. (2022). Crafting public narrative to enable collective action: A pedagogy for leadership development. *Academy of Management Learning and Education*. https://doi.org/10.5465/amle.2020.0224.

Greene, M. (2008). Commentary: Education and the arts: The windows of imagination. *Learning Landscapes, 1*(3), 17–21.

Ilmola-Sheppard, L., Rautiainen, P., Westerlund, H., Lehikoinen, K., Karttunen, S., Juntunen, M.-L., & Anttila, E. (2021). *ArtsEqual: Equality as the future path for the arts and arts education services*. CERADA. https://urn.fi/URN:ISBN:978-952-353-043-0.

Jackson, P. (1968). *Life in classrooms*. Rinehart and Winston.

Laes, T., & Westerlund, H. (2018). Performing disability in music teacher education: Moving beyond inclusion through expanded professionalism. *International Journal of Music Education, 36*(1), 34–46. https://doi.org/10.1177/0255761417703782.

Maris, V. (2015). Back to the Holocene: A conceptual, and possibly practical, return to a nature not intended for humans. In C. Hamilton, F. Gemenne, & C. Bonneuil (Eds.), *The Anthropocene and the global environmental crisis: Rethinking modernity in a new epoch* (pp. 123–133). Taylor and Francis.

Morton, T. (2018). *Being ecological*. The MIT Press.

Nikkanen, H. M., & Westerlund, H. (2017). More than just music—Reconsidering the educational value of music in school rituals. *Philosophy of Music Education Review, 25*(2), 112–127.

Sandlin, B. (2009). Understanding, mapping, and exploring the terrain of public pedagogy. In J. A. Sandlin, B. D. Schultz, & J. Burdick (Eds.) *Handbook of public pedagogy* (1st ed., pp. 1–6). Routledge.

Schön, D. (1983/2016). *The reflective practitioner: How professionals think in action*. Routledge.

Schubert, W. H. (2010). Outside curricula and public pedagogy. In J. A: Sandlin, B. D. Schultz, & J. Burdick (Eds.). *Handbook of public pedagogy: Education and learning beyond schooling* (pp. 10–19). Taylor and Francis.

Schuermans, N., Loopmans, M., & Vandenabeele, J. (2012). Public space, public art and public pedagogy. *Social and Cultural Geography, 13*(7), 675–682.

Senge, P. M., Smith, B., Kruschwitz, N., Laur, J., & Schley, S. (2008). *The necessary revolution: How individuals and organizations are working together to create a sustainable world*. Doubleday.

Senge, P. M. (1990). *The fifth discipline: The art and practice of the learning organization*. Doubleday.

Sjöstedt, M. (2015). Resilience revisited: Taking institutional theory seriously. *Ecology and Society, 20*(4). https://www.ecologyandsociety.org/vol20/iss4/art23/.

Small, C. (1999). Musicking—The meanings of performing and listening. A Lecture. *Music Education Research, 1*(1), 9–21.

Solbrekke, T. D., & Sugrue, T. (2011). Professional responsibility—Back to the future. In C. Sugrue & T. D. Solbrekke (Eds.), *Professional responsibility. New horizons of praxis* (pp. 11–28). Routledge.

Tett, G. (2015). *The silo effect. Why every organisation needs to disrupt itself to survive*. Abacus.

Tran, L. T., & Nguyen, N. T. (2015). Re-imagining teachers' identity and professionalism under the condition of international education. *Teachers and Teaching: Theory and Practice, 21*(8), 958–973. https://doi.org/10.1080/13540602.2015.1005866.

Ulrich, W., & Reynolds, M. (2020). Critical systems heuristics: The idea and practice of boundary critique. In M. Reynolds & S. Holwell (Eds.), *Systems approaches to making change: A practical guide* (pp. 255–306). Springer.

Waddock, S. (2021). Art, Leadership, and System Transformation. Academia Letters, Article 834. https://doi.org/10.20935/AL834.

Westerlund, H., & Partti, H. (2018). A cosmopolitan culture-bearer as activist: Striving for gender inclusion in Nepali music education. *International Journal of Music Education, 36*(4), 533–546. https://doi.org/10.1177/0255761418771094.

Westerlund, H., Karlsen, S., & Kallio, A. (2021). Professional reflexivity and the paradox of freedom. Negotiating professional boundaries in a Jewish ultra-orthodox female music teacher education program. *International Journal of Music Education, 39*(4), 424–437. https://doi.org/10.1177/0255761421988924.

Westerlund, H., Karttunen, S., Lehikoinen, K., Laes, T., Väkevä, L., & Anttila, E. (2021). Expanding professional responsibility in arts education: Social innovations paving the way for systems reflexivity. *International Journal of Education & the Arts, 22*(8), https://doi.org/10.26209/ijea22n8.

Woodford, P. (2019). *Music education in an age of virtuality and post-truth*. Routledge.

SPRINGER NATURE

GPSR Compliance

The European Union's (EU) General Product Safety Regulation (GPSR) is a set of rules that requires consumer products to be safe and our obligations to ensure this.

If you have any concerns about our products, you can contact us on ProductSafety@springernature.com

In case Publisher is established outside the EU, the EU authorized representative is:

Springer Nature Customer Service Center GmbH
Europaplatz 3
69115 Heidelberg, Germany